From the doctor's caseboo

Mrs F. G. weighed 194 pou~~~~~~~~~~~~ ~~~ ~~~ ~~~~ 5 ft 1 in. She had tried to reduce fourteen different times without success. The doctor's quick weight loss diet helped her lose 10 pounds the first week. Within a year she was at 105 pounds, which she has maintained.

Mr D. E. at thirty-two was 243 pounds instead of 153–170, the ideal weight, and he suffered from high blood pressure. the diet put him down to 183 pounds in eleven weeks. He'll stabilize at about 165 pounds. His blood pressure has dropped and he feels like a new man.

Mrs E. F. weighed 155 pounds instead of her ideal weight of under 100 pounds for her 5 ft 2 in. In two months on the doctor's quick weight loss diet she was down to 115 pounds.

During his fifty years of experience in the practice of medicine, Irwin Maxwell Stillman, MD, has helped over 10,000 overweight men and women to reduce. A Diplomate in Internal Medicine, Dr Stillman is presently a consultant in Internal Medicine at Coney Island Hospital, Brooklyn, New York.

Irwin Maxwell Stillman MD, D–IM
Samm Sinclair Baker

The doctor's quick weight loss diet

Pan Books London and Sydney

First published in the United States 1967 by Prentice-Hall Inc
The first British edition published 1970 by Pan Books Ltd,
Cavaye Place, London SW10 9PG
11th printing 1978
© Samm Sinclair Baker and Irwin Maxwell Stillman, MD 1967
ISBN 0 330 02439 6
Made and printed in Great Britain by
Cox & Wyman Ltd, London, Reading and Fakenham

Dedicated with love
to my wife Ruth
IMS

CONTENTS

1. The Quick Weight Loss Diet Succeeds Where Others Fail

The prime aim of this book is to *take that weight off you quickly*, and then to help you stay slim, healthy, and attractive. Here you'll learn exactly how in clear, simple, *proved* ways never told to you before.

You can achieve your desired weight reduction, no matter how many times you may have failed up to now. That's because you'll be using the Quick Weight Loss Diet and other speedy, safe and effective reducing procedures I have developed in treating over 10,000 overweight men and women in more than forty-five years of medical practice.

The necessary, workable instructions that have helped my patients are condensed for you on these pages. All essential details are included. Your special advantage is having the directions handy always for repeated reference and checking.

Once down to your ideal size and figure, you're not left to fight weight increase on your own. As my patients have done with triumphant success, you follow through with the Quick Loss and Stay Slim Eating Combination. You'll find it as easy and remarkably effective as they have. It can work wonderfully for you for the rest of your healthier, more vigorous life.

There are specific built-in reasons why the Quick Weight Loss Diet has succeeded with many hard-to-reduce women and men who had just about given up hope of ever being slim again. Here are the most important:

1. Much more of the body fat is burned off with the Quick Weight Loss Diet than with most other diets, including the most rigorous and difficult. Eating this way you consume fewer calories each day; also your system burns up about 275 *more*

calories daily than with a diet of the same total calorie intake that includes other foods.

2. You don't count calories on the Quick Weight Loss Diet, although calories *do* count. What happens is that you're taking in fewer calories daily without keeping track of them. As you eat, you find that your food consumption is self-limiting. You eat plenty but you don't 'hunger' for more than will take off weight quickly.

3. You lose weight rapidly with the Quick Weight Loss Diet which alone is a great aid in continuing the diet.

4. It's a relatively easy reducing method to follow when you're dining out in restaurants. This has proved to be a crucial asset for busy overweight men and women among my patients. You are permitted just about as many helpful 'coffee breaks' as you want, within the dieting method.

5. At intervals, as many of my patients do, you return to the Quick Weight Loss Diet if necessary for only a few days or a week to keep extra pounds from 'creeping up' on you. The unique combination of intermittent Quick Loss and Stay Slim Eating, as given later in detail, will help you once and for all to take off poundage and then stay slim thereafter.

Only this dieting method helped me bring down my weight from 190 to 140 pounds quickly. I would have been dead years ago if I hadn't taken off that excess fat in a hurry. The same is true of many who have used this diet. This includes doctors and their wives and families, nurses, dieticians, life insurance experts, actors, models, and others who came to me because they had failed to reduce successfully with either various fads and gimmicks or on 'regular reducing diets'.

In one case, a woman who had 'tried everything' without being able to reduce, dropped 25 pounds the first week on the Quick Weight Loss Diet, 67 pounds in four months. She is now on the highly-effective Combination Plan of Stay Slim Eating and occasional Quick Loss dieting which you will learn to apply for yourself. She is maintaining her ideal weight, has improved her health and has achieved a lovely slim figure, to the amazement and delight of her family and friends.

Hers is the largest weight loss in one week of any patient on my records, but a long-range loss is typical. Neither she nor any other of my patients, not a single one, suffered ill effects from quick and dramatic weight losses. Every one of them enjoyed a new glow of healthy vigour and vitality compared to their previous state. And each undoubtedly added years of life – buoyant, healthful years.

It is most satisfying and encouraging to a dieter to see the weight on the scale drop rapidly and the fat 'melt away', as happens with this method. One patient who lost 8½ pounds in her first week on the diet told me, 'After failing on many diets, the big loss the first week has given me the confidence I needed. I was never able to lose so much before. Now I know I can lose weight and I'll stick with it until I reach my goal of a youthful figure that will make people take notice.'

You too will be so delighted and impressed by seeing the decided drop in weight when you step on the scale each morning, that you'll keep dieting another day, then another and another. Dieting 'just one more day' is a key to success in losing weight. The aphorism applies that 'nothing succeeds like success'.

It's an unfortunate fact of the usually recommended slow 'balanced dieting' that when a person sees his weight dropping only a pound or so a week, he invariably says, 'What's the use of even trying? I'll never make it.' And he goes back to stuffing himself again. But my records show that my considerably overweight patients averaged a loss of 6 to 7 pounds in the first three days on the Quick Weight Loss Diet, almost as dramatic as the loss on a fasting diet. Such impressive evidence that it worked, seeing results on the scale with their own eyes, kept them on the simple, plentiful, satisfying diet.

This method will take off excess weight for 95 out of every 100 persons. If you're one of the normal 95 per cent, and you follow this diet and still don't lose weight rapidly but keep on gaining weight, *then you're lying to yourself and to others*.

It's a generally accepted estimate that only 5 out of every 100 persons have some metabolic disorder which keeps them from losing weight even on a rigid reducing schedule, only 5 per cent of the entire population. If you're one of those rare 1-in-20 individuals, your doctor undoubtedly told you so on your last examination. If you have no such complicating physical disorder, this method will take off your excess weight; just follow the brief, uncomplicated instructions.

The situation that is overwhelmingly true of most over-weight people is that you're kidding yourself if you think that some mysterious ailment would prevent you from reducing on this or any other diet. As a doctor advised a fat patient, 'The tests show that your thyroid is normal, your problem is an over-active fork.'

The Combination Plan solves another problem voiced by most dieters: 'Sure, a stringent diet will take off weight for a little while, but then it comes right back. Whatever I lose, I'll gain back again – and more!' As a result of this prevalent conviction, you may have been so sure in the past that you couldn't succeed that you don't even start any more. But regaining lost weight need not happen, as thousands of my patients can tell you.

Using the Quick Weight Loss method combined with Stay Slim Eating, most of these men and women of all sizes, weights, and ages have lost their dreadful excess poundage and have continued to maintain their ideal reduced weight according to my 'Ideal Weight Table' (see Chapter 3).

This method acts to change your harmful overeating habits and to break the custom of gorging at meals and snacking continually on high calorie foods. In this way alone, by stopping vicious, harmful eating cycles, the diet rules help you to reduce and stay slim *after* the dieting period is past. Then the Combination Plan takes over to help you keep your weight down month after month and year after year.

Don't let the melancholy derogators of anything 'new' discourage and deter you from starting the Quick Weight Loss Diet. Hundreds, thousands of overweight individuals have come to me with pleas like this: 'Doctor, I've tried to cut down on my eating and calories on a "balanced diet" but I just can't. I've failed dozens of times. Can you help me?'

According to those who advocate only 'regular balanced dieting', I should answer, 'I can't help you if you won't regulate your eating to limit yourself to smaller portions only of the foods on a "balanced diet" list.' But I know that most of them would fail again with such advice, like the all-too-typical fat lady in a cartoon who told a waiter, 'I'll have a cup of bouillon, cottage cheese salad, melba toast, clear tea and heaven help me, a banana split.' Another moaned, 'I went on one of those balanced diets for four weeks, but all I lost was twenty-eight days.'

The stark truth which must be faced by all, the medical profession and the public alike, is that most people who are overweight love to eat though they may deny it. When told to go on a reducing diet which usually consists primarily of eating smaller portions of a wide variety of foods, most of them just can't limit themselves. Even if they start, they can't stick to it. So they never really even get started.

Your Greatest Danger – and How to Overcome It

Back in 1959, Dr Louis M. Orr, then president of the American Medical Association, was asked in a newspaper interview, 'Do you consider cancer as the greatest threat we face?' He answered, 'No. Cancer is the most dreaded disease in the United States. But the greatest danger to the health of the American people is obesity.'

Now, years later, in spite of such warnings, tens of millions of Americans are overweight. They're not taking advantage of the quick weight loss techniques that work.

I'm constantly appalled by the ingrained conviction most fat people have, and too many wives, husbands, relatives, and friends of overweights, that to change one's habits and eat 'differently' is a harmful thing to do. Consider sensibly and calmly the illogic of this attitude: the way you've been eating has made you fat – *that's the way that's bad for you.*

Almost any change in the direction of consuming fewer

calories must be a change for the better. And obviously the sooner you change, and usually the more drastically you change, the healthier you'll be, and the quicker.

Generalized advice which is usually fed to you, no matter how good it may be, seldom works for you. By following the specific dieting recommendations here you'll lose excess weight, continue on healthful stay slim eating, and never become a fatty again.

In my experience there's no question that quick-action dieting which alters radically and suddenly your ways of eating is the best way to reduce and enjoy the many healthy benefits that follow. Most overweights need the shock treatment of having their eating clearly defined by a specific diet plan which takes pounds off speedily and dramatically. That demonstrated proof of the promise is essential in order for most dieters to succeed. You then go on eagerly to lose the dangerous excess weight that it is urgent for you to take off for your health's sake.

With all its acknowledged benefits, being slim is not a cure-all in itself. Being at ideal weight is not a guarantee against all illness or of extra long life. But your chances of staying healthy or becoming healthier, of avoiding a heart attack and many other serious ailments, of living a longer, more vigorous lifetime, are increased enormously when you melt away layers and pockets of fat that burden your organs.

What's more, and this is of extreme importance in every facet of social and business life, you'll certainly look more attractive to yourself and to others. Most of my patients have accomplished all this safely and happily after other methods had failed for them, by using the quick loss dieting techniques detailed in this book.

I created the Quick Weight Loss Diet and the Combination Plan because my overweight patients needed this specific help. They had never succeeded in reducing on 'regular balanced dieting'.

I urge you to get started now on this diet. Every day that you delay and keep or increase that excess poundage is impairing your appearance and your health, and is endangering and shortening your life. Overweight is a kind of death for almost everyone, slow for some, quicker for others.

When you have got down to your desired ideal weight and figure with this method, have a checkup by your doctor if you didn't consult him at the start. This will be valuable reassurance to you that not only didn't the dieting impair your health, but that you're in far better condition in every way.

Of course you'll know this already thanks to the wonderful difference in the way you feel and from the compliments you get on your improved appearance.

Analysis of Thirty Overweight Patients

While writing this book I analysed again the records of thirty of my most recent overweight patients. This special little study revealed interesting and promising facts for you:

First, it was remarkable how quickly and easily they all learned the Quick Weight Loss dieting technique and how uniformly it kept each one dieting and losing pounds (they dieted as individuals, not as a group).

Second, on the occasions when some of them slipped off the Quick Weight Loss Diet, they knew from experience that a return to it would take weight off rapidly. Each one went right back on the quick-reducing instructions.

Third, their records (and thousands of others) proved the fallacy of a too-prevalent viewpoint that a fast take-off of weight is linked inevitably with a fast put-on. This is not true. I have seen too many people struggle with slow-take-off diets, then slip and put on weight quickly – and give up. They don't know how to take off those added pounds in a hurry so they're discouraged from trying again. This does not normally happen with the Quick Weight Loss Diet because the dieter soon is aware of the fact that his fat *can* come off in a hurry – *it has happened to him.*

2. How The Quick Weight Loss Method Works

The Quick Weight Loss Diet has built into it elements to reduce you even by subterfuge. For example, there is no in-
struction to count calories. You can eat as much of the fine
foods on the list as you need to satisfy your hunger: a great
variety of lean meats, unfatty fish and shellfish, chicken and
turkey, eggs, and cottage cheese.

Your first reaction may well be that of so many of my
patients, 'But on that diet I'll eat my head off. I'll consume far
more calories than I do now.' The fact is that on this diet, even
eating six or more meals a day, you'll find that you limit the
intake yourself because you feel full and lack the desire for
excess quantities that would increase your weight instead of
taking it off. Also you're restricted from the rich foods that
tempt you to overeat. And you eat to satisfy your hunger, not
to pamper your appetite. A very few overeat for the first day
or two but then taper off quickly.

Many of my patients have told me, 'I feel I'm eating so
much on that diet, how am I losing so much weight?' When I
added up their daily intake I found that they had consumed
as few as 400 to 900 calories a day, without any desire for
more food. Others, primarily taller persons, were eating up
to 2,000 calories a day on the Quick Weight Loss Diet and yet
lost pounds rapidly although they hadn't been able to lose on
other diets where they seemed to be eating less. The high-
protein consumption burned up many more calories per
day.

With this method, the time comes eventually when you are
either at or near your ideal weight. For the average person who
is reducing drastically, it is best to come within 5 to 15 pounds
of the ideal weight, then go on Stay Slim Eating for a few

months. After that you take off the remaining 5 to 15 excess
pounds through the Quick Weight Loss Diet, then back to
Stay Slim Eating when your ideal weight is achieved.

Here are just a very few typical records of how individuals
reduced with the aid of this method. In every one of these
successful cases and thousands more, not one patient showed
any harmful effects. Every one improved his or her health and
well-being by losing excess weight. The usual comment:
'Doctor, I feel wonderful. I have new energy. Everything looks
brighter. My husband and friends tell me that I look years
younger and far more attractive.'

One of my patients is a physician who was considerably
overweight. She is 5 ft 3½ ins. and weighed 183 pounds. I placed
her on the Quick Weight Loss Diet, prescribed medication and
told her to eat only if hungry. She never stopped her practice
as a busy doctor. On her own decision, she went for seven
days on water and vitamins, starting on my diet only from the
eighth day. She lost 13 pounds the first week, 12 pounds the
second week, 6 pounds the third week – dropping from 183 to
152 pounds, 31 pounds in three weeks. She will continue to diet
until she stabilizes at 125 pounds, then will go down eventually
to 110 pounds, about her ideal weight. Her opinion of the
Quick Weight Loss Diet: 'It's great!'

Quick Weight Loss for Families

Entire families lose pounds wonderfully through the Quick
Weight Loss Diet (it usually helps when two or more people
go on the diet together and, in effect, have a contest with each
other). In one instance of a family of four, during the first four
days on the diet the considerably overweight father dropped
9 pounds. Mother, only a few pounds overweight after a heavy-
eating cruise, lost the four pounds she'd gained on vacation.
The fourteen-year-old overweight daughter dropped 6 pounds.
Even the average-weight sixteen-year-old son who was not on
the diet itself went down a pound, influenced by the way the
others were eating.

They were all so delighted that the two overweights continued dieting and losing weight. The happiest result was that the father, who had never been able to stick with a diet before, reduced to a slim, attractive figure his children had never seen. He'd given up that goal many years before as unattainable for the rest of his life. One doctor after another had told him to reduce for his health's sake but he hadn't succeeded.

The family called the diet a 'miracle'. Actually it was just another proof of the sound functioning of the quick reducing method. People need specific instructions that work, not outmoded reducing theories amounting to a recommendation to 'eat less' – which are bound to fail for most.

Some Typical Cases on the Quick Weight Loss Diet

A. B. – man, 5 ft 4 ins. He weighed 170 pounds and despaired of being able to reduce as he had failed many times. I recommended that he come down to 125–130 pounds, ideal weight for him. On my diet, he lost 7 pounds the first week. His rate of loss varied because of lapses in dieting, but he was down to 127 pounds in three months, a loss of 43 pounds and at ideal weight. He felt so good that he had refused to stop dieting on the way down, and he continued to cut down for another six weeks until he weighed 116 pounds.

I emphasized that there was no advantage to him in being underweight so during the next month he rose to 127 pounds. Gradually, not taking care in following the Combination Plan, over the next four months he went up to 138 pounds. Going back on the Quick Weight Loss Diet which he had learned was relatively easy, he then reduced within a month to 128 pounds where his weight has now been stabilized for some time, as charted.

B. C. – woman, 5 ft 3 ins. She weighed 147 pounds, over 30 pounds heavier than her ideal weight of 103–115. On the Quick Weight Loss Diet, she dropped 6 pounds the first week and was down to 130 pounds in six weeks. I told her to main-

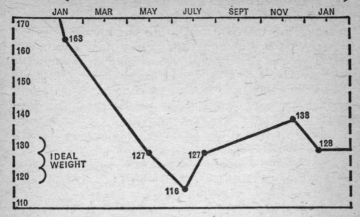

A. B. man 5 ft 4 ins. tall one year record

tain that weight for a few weeks, then go down to 120 and stay there for six months before stabilizing at about 110. We both knew she could do it because she had learned how relatively easy it had been to take off over 25 pounds.

C. C. – woman, 5 ft 5 ins. She weighed 174 pounds, dragging around a load of about 50 pounds more than her ideal weight of under 125. She told me, 'Doctor, I feel terrible in every way. I've tried to diet gradually but it doesn't work for me.' I started her on the Quick Weight Loss Diet and also prescribed a drug for assistance. At the end of a week she had lost 10 pounds and was down to 164.

'I'm feeling so much better already,' she told me, 'that nothing can stop me now from getting down to normal weight. When I lost a few pounds the first week on other diets, it didn't mean anything to me, but 10 pounds – that really proves something. My husband said I'll soon be looking like his bride again.' Eventually she reached her ideal weight.

D. E. – man, 5 ft 11 ins. He weighed 243 pounds instead of between 153 and 170 pounds ideal weight. He had high blood pressure and many other undesirable symptoms and ills. He said he felt like an old man although he was only thirty-two. On my diet he went down to 183 pounds over a period of fourteen weeks. His blood pressure dropped from 210/135 to 150/100. He will be on Stay Slim Eating for a while and then will lose 20 to 25 pounds, stabilizing at about 165. With the big weight loss already, he says that he feels like 'a new man'.

E. F. – woman, 5 ft 2 ins. She weighed 155 pounds instead of her ideal weight of under 110. She said she was determined to take off weight because she heard someone refer to her as a 'butterball'. In two months on the Quick Weight Loss Diet she was down to 115 pounds. Letting up too much, her weight rose to 124 pounds in a month. On the Quick Loss Diet she dropped to 115 pounds in two weeks and then to 110.

She asked me, 'Will I have to be careful about how much I eat for the rest of my life?' 'Yes,' I told her, 'just like millions of others who are anxious to stay thin, healthy, vigorous, and attractive. Isn't it worth the difference in the way you look and feel now?' She agreed that it was. She is proving it by maintaining her ideal weight on the Combination Plan, returning to the Quick Loss Diet for brief periods whenever her weight rises, which is seldom.

F.G. – woman, 5 ft 1 in. She weighed 194 pounds and said that she had tried to reduce *fourteen* different times without success. Each time she had lost a few pounds and then gained the weight right back again. By the Quick Loss method, she lost over 10 pounds the first week. She was so encouraged that she continued eagerly. Within a year she was at her desired weight of under 105 pounds and has maintained this for over four years at this writing.

In typical cases like the preceding, the loss of weight takes dangerous strain off the heart. I'm often asked, 'Doesn't such dieting put a great strain on the heart due to the sudden change?' The answer is precisely the opposite; taking off excess fat takes strain off the heart. The electrocardiograms (on pages 23–4) of an average individual show how this becomes apparent.

With the aid of quick loss dieting, years of life were undoubtedly added for this patient. Dangerous heart impairment from overweight was avoided. The fat crowding the organs was eliminated. The organs returned to normal and the signs of heart strain disappeared.

3. The Goal: Your Ideal Weight

The Quick Weight Loss Diet will help you attain your ideal weight quickly. This 'ideal weight table' based on my findings differs from others. I don't believe that you should be guided by what the average weight is for persons of your height but rather by what your ideal weight should be for maximum health, well-being, and long life. The average weights listed by others are often too high, in my opinion, because too many Americans are overweight and this raises the averages above the ideal weight.

I don't give figures by 'type of frame' because this is misleading to many. What comprises a small, medium, or large frame isn't easy for the individual to decide for himself. Nearly all my overweight patients have claimed to have large bone structure although only a small percentage were in that category. Very few chose to consider themselves as having a small frame because it would mean that their weight should be even lower.

There are too many variations for you to be able to determine correctly what kind of frame and bone structure category fits you. Many people have broad shoulders and a small pelvis. Others have just the opposite. Some have a thin chest and heavy legs and thighs. Some have short, thick necks but with thin hands and arms, and large feet and legs – or just the opposite.

I recommend that you aim for your weight listed on the following weight and daily calorie table, a few pounds one way or the other; lower is better than higher. The lower figure is the 'ideal weight' (which may be called 'desirable weight' by some statisticians or doctors). The higher figure is the 'average weight' for the height. These figures are for men and women aged twenty-five and over, in stockinged feet, and without jackets. Note also your daily calorie intake to maintain ideal weight.

(1) The first simplified graph depicts an electrocardiogram at normal weight with what is known as the T-wave rounded or peaked as here:

(2) As the person gains weight, the T-wave is slightly flattened:

(3) With further gain of weight, the T-wave has disappeared, a clear warning that the heart strain is imminent:

(4) The person has now put on additional weight, is considerably overweight and the T-wave is inverted, a certain sign of heart strain that can be serious:

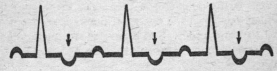

(5) Fortunately the individual was warned to diet or possibly incur serious heart damage. As the weight came off, the T-wave direction was reversed and the inverted line flattened:

(6) With further loss, the T-wave peak began to appear again:

7) With successful dieting, at desired weight, the T-wave is peaked again, the danger signal of heart strain gone:

WEIGHT AND DAILY CALORIE TABLE

Women		Men		
Weight	Daily Calorie Intake	Height	Weight	Daily Calorie Intake
95–100	1,200–1,500	5 ft	105–110	1,320–1,650
95–105	1,260–1,575	5 ft 1 in	105–115	1,390–1,735
100–110	1,320–1,650	5 ft 2 ins	110–121	1,450–1,815
103–115	1,380–1,725	5 ft 3 ins	113–126	1,520–1,900
108–120	1,440–1,800	5 ft 4 ins	120–132	1,585–1,980
112–125	1,500–1,875	5 ft 5 ins	123–137	1,650–2,065
117–130	1,560–1,950	5 ft 6 ins	130–143	1,715–2,145
121–135	1,620–2,025	5 ft 7 ins	133–148	1,780–2,230
124–140	1,680–2,100	5 ft 8 ins	140–154	1,850–2,310
130–145	1,740–2,175	5 ft 9 ins	143–159	1,915–2,400
135–150	1,800–2,250	5 ft 10 ins	150–165	1,980–2,475
140–155	1,860–2,325	5 ft 11 ins	153–170	2,045–2,560
144–160	1,920–2,400	6 ft	160–176	2,110–2,640
148–165	1,980–2,475	6 ft 1in	163–181	2,180–2,725
153–170	2,040–2,500	6 ft 2 ins	170–187	2,245–2,800

The preceding weight table is figured generally as follows:
WOMEN aged twenty-five or over: 5 feet – 100 pounds.
For every inch above 5 feet, add 5 pounds per inch, for aver-

age weight; from that figure subtract about 10 per cent for ideal weight; the figures give you the range between ideal and average weight as a guide. For eighteen to twenty-five years of age, subtract 1 pound for each year under twenty-five. For example, a girl of nineteen, 5 ft 4 ins. tall, ideal weight – 102 pounds, average weight – 114 pounds.

MEN aged twenty-five or over: 5 feet – 110 pounds. For every inch over 5 feet, add 5½ pounds per inch for average weight; from that figure subtract about 10 per cent for ideal weight; the figures give you the range between ideal and average weight as a guide. For eighteen to twenty-five years of age, subtract 1 pound for each year under twenty-five. For example, a young man of twenty-one, 5 ft 10 ins. tall, ideal weight – 146 pounds, average weight – 161 pounds.

ADDED SUGGESTIONS: If you're a professional in the theatrical or entertainment fields, a clothes model, an athlete or a dancer, it is best that your weight be about 5 pounds less than the ideal weights listed.

If you have been considerably overweight for years I recommend that you reduce with the Quick Weight Loss Diet to within 5 to 15 pounds of the average weight for your height. After maintaining that figure for six months to a year, I suggest that you reduce further to your ideal weight and maintain it thereafter.

CALORIE INTAKE: If you wish to calculate your daily calorie intake according to the average weight for your height, figure 12 to 15 calories per pound.

For example, if the average weight for your height according to the preceding table is 120 pounds: $120 \times 12 = 1,440$ calories; $120 \times 15 = 1,800$ calories. Your calorie intake to maintain that weight would be 1,440 to 1,800 daily. In reducing, you add up everything and eat 1,440 to 1,800 calories daily (if you're overweight you're now consuming many more calories than that). But this type of attempted self-discipline which results usually in slow weight loss rarely works.

Ineffective Balanced Dieting is Useless

An advertisement for meats and poultry was headlined, 'The Secret of Weight Control: Cutting Down – Not Out!' It went on to state, 'You don't need to be scientific about what you eat, just sensible ... and forget about fads. Eat a variety of foods and not too many calories, that's the way to good nutrition. Not a very complicated idea, but a sound one!'

All that is true as far as it goes, which is far short of the reducer's goal. Balanced eating is a very 'sensible' system. But for most overweights it just doesn't work. And if it doesn't work for you then the concept of balanced dieting is useless for you because it doesn't take off your excess weight.

There's no question in the minds of these people of *wanting* to be slim. The question is how you can accomplish it realistically and surely. Hundreds, thousands of persons have told me, 'Doctor, I've tried to cut down on my eating and on calories but I just can't. I've failed dozens of times.' Am I, as a doctor, then supposed to fail my patients?

According to those who advocate balanced eating only, one should say in effect, 'I can't help you if you won't regulate your eating and limit yourself to smaller portions of only the foods on a balanced diet list.' But I know that most overweights will fail on such advice just as surely as if I said, 'No special diet, just balanced eating – and buy your clothes two sizes too big.'

I've had to find out over the years, by trial and error, what will help most overweights to reduce, not just what they should do but what they *will* do. Accordingly I've devised these quick loss diets, methods and specific instructions and recommendations. My experience has proved that these techniques work. The pounds come off safely, swiftly, and far more surely with the Quick Weight Loss Diet and with the many bizarre quick action diets listed later to accommodate your personal needs.

Get Started Now ...

I recommend that you start at once with the proved Quick
Weight Loss method on which *you don't count calories*. You
will lose weight quickly because of the restricted types of foods
and other elements which result in your consuming fewer calories and *burning up more calories* than with varied eating.

Please keep this in mind every single day that you're reducing: you are not guessing that the Quick Weight Loss Diet
will work. You are not relying on false claims or empty pro-
mises. This method has been proved in the cases of many
thousands of others who have regained attractive figures. It
will work for you – swiftly, pleasantly, surely. Just give the
Quick Weight Loss Diet at least one week's trial. Never again
will you moan hopelessly in Shakespeare's words: 'O! that
this too too solid flesh would melt.' This time you can have
confidence that you'll achieve the wonderful reducing results
and the slim figure you want.

4. The Proved Quick Weight Loss Method

Not only from repeated successes with thousands of men and women, but also from extensive study and analysis of every kind of diet I could find, I'm convinced that the Quick Weight Loss Diet is the finest and best for most people who should take off excess weight. Judging from the records of my patients, if you adhere to this simple, satisfying diet faithfully, *you should expect to lose 5 per cent to 10 per cent of your weight the first week if you are overweight*. If you weigh 150 pounds and are considerably overweight, you must lose between 7 and 15 pounds the first week.

Another way of figuring your probable loss according to the averages is that you should lose 5 pounds a week, week after week. If you are 10 pounds overweight, it should take you two weeks to lose that excess. If you're 15 pounds overweight, you'll probably take off that excess weight in only three weeks.

There may be weeks when you'll lose 3 pounds or so, and other weeks when you'll drop about 10 pounds. It's not unusual for a considerably overweight person to lose 15 to 25 pounds the first week (25 pounds being about the greatest loss in a week on my records with the Quick Weight Loss Diet, as previously noted).

You score these satisfying losses while you eat excellent wholesome foods on this diet – lean meats, fish, poultry, shellfish, eggs, and cottage cheese. You can have as much coffee and tea as desired, without sugar or cream or milk but with non-caloric sweeteners if you like. You may drink as many non-caloric carbonated beverages as you wish.

You'll note definite 'must' restrictions on the diet but most of my patients have not found them too limiting or difficult. Right from the start you take off excess fat quickly to reach your desired weight. Then you may indulge yourself in the foods you enjoy most, within reason. It's important that you weigh yourself every morning. If you see that you're 3 pounds or more above your ideal weight, back you go on the Quick Loss Diet for a week to burn off unwanted fat. Each time you go back on the diet you'll find it easier to maintain your ideal weight afterwards.

The Quick Loss Method

Eat all you wish of the following foods to satisfy hunger. If you eat until you've satisfied your hunger but haven't stuffed yourself, and then feel hungry before the next meal, you can eat again but only from the foods listed here. In fact, you're better off eating smaller meals six times a day than three bigger meals as is the general custom. But you limit yourself to this extensive selection of high protein foods:

1. LEAN MEATS, with all possible fat trimmed off. Includes beef, lamb, veal. Eat it grilled, boiled, baked, or smoked. No butter, margarine, oil, or other fats or grease are to be used either in the cooking or in the eating.

2. CHICKEN AND TURKEY with all skin removed. Grilled boiled, or roasted, with no butter, margarine, oil, or other fats or greases. Young fowl is preferable but not necessary.

3. ALL LEAN FISH such as flounder, haddock, cod, perch; SHELLFISH: shrimps, scallops, lobsters, oysters, clams, crabs. Grilled, baked, or boiled. No butter, margarine, oil, or other fats or greases permitted in preparation or in serving. You may have cocktail sauce, horseradish, ketchup – in moderation.

4. EGGS. Hard-boiled, preferably. Also soft- or medium-boiled eggs, poached, or any type of fried eggs made in non-fat, non-stick pans without butter, margarine, oils or other fats or greases.

5. COTTAGE CHEESE and other cheeses made with skim milk and no whole milk.

6. AT LEAST EIGHT GLASSES OF WATER DAILY, absolutely essential (10-ounce glasses). As much as you want of

coffee and tea without cream, milk, or sugar. You may sweeten with non-caloric sugar substitutes to taste. Enjoy all the club soda, vichy, and non-caloric carbonated beverages you wish, as often as desired.

The eight glasses of water daily are in addition to whatever coffee, tea, and non-caloric carbonated beverages you drink. This is an integral part of the specific internal process put into motion by the Quick Weight Loss Diet. Much more of your body fat is burned up than with most other diets. This leaves in its wake the waste products or ashes of burnt fat. They must all be washed out of your system by water which also serves to relieve any unpleasant dryness and taste in the mouth.

Any time you feel a great desire for sweets, pour a glassful of non-caloric carbonated beverage, varying the flavours you like most. This is a treat you may enjoy just about as many times a day as you wish. Such drinks are refreshing and satisfying to your 'sweet tooth'. They contain no carbohydrates and add to the desired flow of liquids through your system.

You may use common seasonings such as salt, pepper, garlic, cloves, thyme, other herbs and spices, cocktail sauce, tabasco sauce, horseradish, ketchup. Don't use any creamy or oily sauces or dressings. No mayonnaise, no salad dressings, no oils, no fats.

It is desirable to take vitamins daily with the diet, such as a once-daily vitamin and mineral tablet which contains vitamins A (5,000 USP units), D (500 units), B-1 (3 mg), B-2 (3 mg), C (50 mg), and additional varied vitamins and minerals. (I don't recommend any particular brand or this formula specifically – almost any once-daily vitamin will do, with varying contents of vitamins and minerals, as offered by many different companies.) If you're very much determined to lose weight rapidly, and you cut down your eating to a minimum on the Quick Loss Diet, take two of the once-daily vitamin tablets each day, or choose commonly available 'therapeutic' tablets or capsules containing higher dosages.

NOTHING ELSE IS PERMITTED ON THIS DIET –

NOTHING! IF IT'S NOT MENTIONED IN THE PRECEDING LIST, DON'T EAT IT OR DRINK IT. The addition of any other food or drink will prevent the full and efficient functioning of the high protein internal reducing process which burns up more fat.

You'll find that this is a relatively easy diet to maintain when you eat at restaurants. For example, at a restaurant you can always get grilled, roasted, or boiled meat, chicken, or fish, shellfish, boiled or poached eggs, cottage cheese. You might start with a shrimp cocktail, complete with cocktail sauce. Simply skip the side dishes and eat as much steak or lean meat as you wish, for instance, to satisfy your hunger – then stop.

If you're dining at someone's home, carry your Quick Weight Loss Diet listing and show it to your hostess. She should certainly honour your need and resolve to take off dangerous excess weight and not be offended that you don't eat her rich desserts or other foods not permitted on the diet.

Like so many others on this diet, you'll probably be pleased to find that soon you don't miss the butter, margarine, oil, or other fats commonly used in preparing and eating foods. Rather the majority of those on this method decide even after they go off the diet that they prefer the clean taste of good food itself, in its more natural flavouring, without calorie-heavy fats added. A typical comment is: 'I used to love French fries but now I can't even stand the thought of them. Give me a clean, naturally tasty, unbuttered baked potato instead any time.'

Only Half a Glass of Water Per Hour

Oddly enough, one of the common reactions I get from people when I recommend the Quick Loss Diet is, 'The variety of fine foods sounds wonderful, and all I can eat too, but how can I manage to drink eight glasses of water every day?'

I asked one woman, 'Can you drink half a glass of water each waking hour?'

'Of course! Who can't drink half a glass of water an hour?'

'Or, a glass of water every two hours?'

'Certainly. Nothing difficult about that.'

'Then,' I concluded, 'what's your concern about being able to drink eight glasses of water a day?'

'Oh . . . I didn't understand. The first thing that entered my head was that I'd have to drink eight glasses of water all at one time . . .'

Was that your first reaction too? If so, it's because your approach to any kind of dieting is negative rather than positive. The first step in taking off that fat and flab is to want to very much, so that you'll start this diet at once and stay with it at least a full week. I know three overweight roommates, young women, who started this diet together and lost 14, 15, and 17 pounds respectively the first week.

Figure out right now how much less you'll weigh at the end of one week. Isn't that a worthwhile goal? Practically any obstacles can be overcome if you care enough. Patients who are not near a convenient supply of water at all times, such as taxicab drivers, carry a jug of cold water with them.

Try the Quick Loss method for a week positively, optimistically, not negatively as though you're depriving yourself of anything important. As I told an overweight patient (F.G.), 'You worry about drinking eight glasses of water a day as a great chore. Yet you have no trouble at all stuffing your fat-choked insides each day with dish after dish of rich heavy food, ruining your health and shortening your life.'

She ran to the water tap in a hurry, started on the Quick Loss Diet right there and then. She dropped from 161 to 151, a loss of 10 pounds, the first week. In six weeks she weighed 135 pounds, 26 pounds taken off in that brief period She is headed for her ideal weight of 120 pounds.

Proved Superior by Thousands . . .

One of the best features of this method, as I have learned from its use by thousands of individuals, is the enthusiasm it arouses among those following it. Practically all have found it 'easy to take'. The drop shown on the scale each morning spurs them to stay within the simple specifications instead of giving up as they had on so many diets so many times before.

With this diet you eat all you want from the foods on the list, enough to satisfy your hunger but not in excess just for eating pleasure. This is in effect a 'demand diet' in that you eat whenever your hunger demands. It is also a form of 'nibbler's diet' since you eat just enough to satisfy yourself and only when you're hungry.

You'll find that you're not as tempted to overeat the foods on the Quick Loss Diet as you are by rich, starchy, sweet, calorie-heavy dishes you might stuff yourself with otherwise. You should not overeat at any time. Stop when you've satisfied your hunger and you'll lose weight rapidly.

You don't count calories because there's no need to do so, the diet itself is self-limiting in effect. When you have taken off the desired weight and shift to Stay Slim Eating, it is then advisable to learn the calorie value of foods and restrict yourself to 1,700 to 2,500 calories daily (for men) or 1,300 to 2,200 calories per day (for women).

To sum up, the Quick Loss Diet has helped more people in my care take off up to 50 pounds of excess weight and keep it off than any other dieting programme I know of.

The diet is listed below in brief format. I suggest that you make two copies of it. Pin up one copy of the list in your kitchen so you'll have it there for constant reminder and reference. Carry the other copy in your pocket or purse for reference if you eat out.

THE QUICK WEIGHT LOSS DIET – eat only food on this list:

1. LEAN MEATS – boiled, baked, or grilled – no butter, margarine, oil, or other fats.

2. CHICKEN, TURKEY – all skin removed – grilled, boiled, or roasted – no butter, margarine, oil, fats.

3. LEAN FISH, SHELLFISH – grilled, boiled, or baked. No

butter, margarine, oil, fats. Cocktail sauce, ketchup, horse-radish, all permitted in moderation.

4. EGGS – preferably hard-boiled, but also may be cooked any way not using butter, margarine, oil, fats.

5. COTTAGE CHEESE, SKIM MILK CHEESES.

6. AT LEAST EIGHT GLASSES OF WATER DAILY ESSEN-TIAL! As much as you want of coffee, tea (without cream, milk, sugar), club soda, vichy, non-caloric carbonated beverages. Non-caloric sugar substitutes permitted.

You may use common seasonings such as salt, pepper, spices, cocktail sauce, ketchup, horseradish.

NOTHING ELSE PERMITTED – IF IT'S NOT LISTED HERE DON'T EAT OR DRINK IT!

5. How This Diet Burns Up 275 More Calories Daily

If you're concerned only with the 'how' of taking off excess weight rapidly, then you can simply use the Quick Weight Loss Diet and get its benefits. You need not read the details here on 'why' this high protein diet functions so well in the human system to reduce you faster. This information is for those who are interested and who find it helps them to keep dieting when they know such 'whys'.

This explanation is in the most simplified terms, certainly not in complete scientific detail which only scholars would understand – they know all this already. The diet is made up entirely of protein, chief constituent of animal bodies. The ability of protein to produce more energy is called the specific dynamic action of protein.

What is important to you is that this diet, which in essence provides protein only, works more efficiently than other types of diets, from a chemical and metabolic viewpoint, in burning up fat. With this method, the system burns up about 275 more calories daily than a diet of the same total caloric intake but including, for example, fruits, vegetables, butter, margarine, and oil.

Most foods that grow have some protein in them and are able to raise the fires of metabolism to some extent. The degree depends on the amount of protein in the foods. Those listed on the Quick Loss Diet have a higher proportion of protein. The resultant burning capacity is excellent so that you eat and burn up your own fat. Some of this fat (your own body fat) is converted into a small amount of carbohydrate or sugar. The remainder of the fat is converted into energy or fuel. The fat is mainly broken down into fatty acids which provide the body with its heat and energy when no carbohydrates or sugars are ingested. This process is a step by step reaction; with each reaction some energy and heat are slowly given off.

In this process, about 60 per cent of the fat is completely oxidized (united with oxygen) into carbon dioxide and water. A large percentage is left as ashes in the 'furnace' to be disposed of. (Unused fatty acids such as oxybutyric acid and aceto-acetic acid are the 'ashes' of fat refuse.)

Studies have shown that persons on a high protein diet such as this, with no carbohydrate intake, melt the fat out of the storage or fat centres. The fat is brought to the blood where it is transported to the liver. Here it is broken down into bile salts and acids. The unused fat is again poured into the blood as ketone bodies.

These are really fatty acids which are brought to the kidneys and carried away in the urine. The body withdraws extra water to dilute these acids. That is why the body not only loses extra fat but also depletes extra intercellular water and is a reason why there is usually more frequent urination with this diet. As a result, the body is lighter, and the heart, lungs, and blood vessels beat, breathe, and flow more easily. Thus most people on the diet feel lighter and better very quickly.

The fatty acids are also slight irritants to the kidney. In order to get rid of them the body must have plenty of water to wash them out. That's why no less than eight glasses of water a day are specified.

If you should take sugars and carbohydrates, here is why it would cut down radically the effectiveness of the Quick Loss Diet. Each gramme (one gramme equals 1/30 of an ounce) of carbohydrate you ingest substitutes for the burning up of 1 to 2 grammes of your body fat, equivalent to 9 or 18 calories. If carbohydrates are taken in, the body ignores using its fat and uses the carbohydrates only. Thus the fat, instead of being burned up, is stored in your fat depots.

The 'why' of this functioning is not fully understood scientifically. But on this regimen, you as a dieter get the advantage of this fact and continue to reduce even though you may be eating permitted protein foods totalling 1,300 to 2,500 calories daily.

If you disregard the instructions and eat regular carbohydrates or sugar or alcohol, you will stop burning up stored fat as efficiently. Even one indiscretion, one mouthful of ice-cream, for example, will set you back a day or two. The diet must be observed strictly or you will not do as well as many thousands before you have done in using up bulging fat.

You will lose the most weight on the Quick Loss Diet, most quickly, if you eat only to satisfy your hunger pangs rather than to gratify a desire to eat just for the pleasure of eating. Understand that with this diet your body supplies the fat and carbohydrates which are lacking in the diet. You use up stored fat for energy and also convert the protein into starches for energy. So the less you eat, the more you use up excess fat stored in your body.

I must make it clear from a professional viewpoint that the Quick Loss Diet is not a 'miracle cure' or a scientific functioning of my invention. High-protein diets are not 'new'; this is my own version, one which I have found to be most usable and effective.

This Quick Weight Loss is Not Mostly Water Loss

Again and again I hear ignorant people, mostly heavyweights who resist dieting, say, 'Quick dieting weight loss is phony – it's mostly water loss. You dump a lot of water the first few days and then the weight comes back.' That is nonsense in respect to my method.

I've had patient after patient lose a great deal of weight the first week and then 5 to 6 pounds consistently week after week on this diet. While some of this weight loss is water, even the worst diehard must admit that most of it is excess body fat.

Consider the case of a patient, a very heavy man, who lost 112 pounds in sixteen weeks on a particularly limited version of this method. He felt fine all the way as he dropped from 354 to 242 pounds. He ate only one meal a day when he felt hungry, at no set hour. He drank a great deal of water and club soda as he put in a long, hard week's work week after week.

He had no ailments or complaints and never felt weak or exhausted. He slept well. His blood pressure which had been much too high dropped to normal. Since he reached his desired weight he has had no difficulty maintaining it.

Here's the week-by-week record of this man's weight loss:

1st week – 17 lb.	5th – 4	9th – 2	13th – 8
2nd ,, – 10	6th – 6	10th – 11	14th – 5
3rd ,, – 6	7th – 6	11th – 4	15th – 6
4th ,, – 7	8th – 13	12th – 1	16th – 6

This loss of 112 pounds certainly was not all water. Nor are the losses by hundreds upon hundreds of individuals whom I've watched drop 40 to 75 pounds and maintain the reduced weight. These people were not waterlogged nor are you unless you have some disorder which your doctor has already noted.

So I suggest that you stop worrying about losing only water. Instead, start Quick Weight Loss dieting. As your clothes become sizes too big for you, you'll agree that you've melted off pounds and inches of all too solid, unhealthy, and unwanted flesh, not water.

What Not to Eat

Your simple guide is to eat only the foods previously listed. Remember, you follow these restrictions for a limited time, not forever. As examples, here are some foods not allowed: no bread, rolls, or cake in any form. No vegetables not even lettuce or celery. No fruits of any kind, no oranges, lemons, or grapefruit in solid form or in juices.

No medical 'fillers' such as methyl cellulose which I'm sure doesn't sound very appetizing to you anyhow.

No alcohol is permitted for the duration of the diet; no brandy, whisky, gin, or even the driest wines or beer. You can have these drinks in moderation, and your favourite foods in moderate portions, after you reach your ideal weight.

No ice-cream, ices, sherbets, or the like, no desserts.

No soft drinks containing sugar. You can drink and enjoy all the non-caloric carbonated beverages you wish, as much as you want. If you take a blindfold test, you won't be able to taste the difference between sodas made with sugar and those made with non-caloric sugar substitutes.

No dairy products outside of the eggs and a few cheeses mentioned in the listing. No whole milk or cream. No sugar in coffee or tea; use as much as you want of the non-caloric sweeteners, in liquid, tablet, powder, or other form.

Don't allow yourself even such seemingly small indulgences as taking a few mint sweets, fruit drops, other small sweets, or chewing gum while you're on this diet. The reason is that you may well be one of the many persons who cannot metabolize carbohydrates properly. Even the relatively small quantity of carbohydrates in a mint or piece of chewing gum seems to be converted into fat in the system rather than burned up for energy and heat. That slight amount disturbs the adjustment in the body whereby the functioning of the Quick Loss Diet burns up fat most efficiently.

The same applies to bread, cake, biscuits, jelly, cereals, noodles, alcoholic beverages, or any of the hundreds of products which contain the slightest amounts of carbohydrates. Don't eat even a tiny biscuit or sip a small alcoholic drink or you'll throw the entire reducing process out of kilter.

In short, if a food is not on the listing, and you question whether or not you're allowed to eat it, the answer is 'No!'

Most of my patients haven't found the restrictions too hard to handle because they realize that they can indulge in such foods and drinks in moderation once they lose their excess weight. They know, as you will, that you go on this diet 'for a limited time only'.

Well-Being on the Quick Loss Diet

The vast majority of people on this diet feel normal and perfectly well physically. Many have told me they felt particularly happy, alert, vigorous, and in high spirits.

For the great majority of persons, the Quick Loss Diet provides all the protein needed, all the fat and vitamins required by the system. There would seem to be a lack of vitamin C on this diet but it's actually provided by the fresh fish, shellfish, and meat. In all my observations, I've never found one person who has developed a clinical vitamin deficiency on this diet. Nevertheless I recommend that you take a once-daily vitamin which supplies the supposedly normal daily vitamin requirement.

Remember that patients of mine have lost as much as 25 pounds the first week and 106 pounds in three months on this diet. Thorough examination of these persons and many others did not disclose any undesirable neurological or physical signs, no vitamin deficiency. It is good medical practice to have your doctor examine you occasionally while on this diet or any diet.

Regarding Waste Elimination

This diet has practically no residue. Therefore it is likely that you may not have a bowel movement for a few days, although not necessarily so. In any case, don't be disturbed by any such lack. If you are at all concerned, take some mineral oil or milk of magnesia. Nature adjusts the bowel habits so that many persons soon have a daily movement again in spite of the lessened residue.

Protein produces a different intestinal flora or bacterial growth. As a result, many patients have told me that they have a new surge of well-being within a surprisingly short time. Many of them quickly lost the bloated abdominal feeling which they'd borne while overweight.

The Quick Loss Diet is also excellent in that the individual usually loses a great deal of water. You'll find that urination increases, often considerably, as water and fats and waste are removed from the body. I've had some patients on this diet lose over 10 pounds of water in a week. *You are also losing considerable fat at the same time.*

Women who retain water on this diet and do not give it up readily usually are about to have a period, or seem to retain water for some unknown reason. If they are among the rare few who have low protein in the blood, they are already aware of it. They should diet under the direction of a doctor. He will prescribe one of many diuretics which will be of assistance in such cases. Eliminating salt from the diet also helps those who retain water.

Additional Points

Since you are an individual and no two persons are exactly alike, you may apply variations of the diet as noted here.

If you're allergic to eggs don't eat any – it's unnecessary. The listing shows what you're permitted to eat, not what you must eat.

If your cholesterol count is high (as your doctor would have informed you in the past), don't eat more than four eggs a week.

If you're allergic to shellfish, omit only those species which affect you adversely. Eat the others, also lean fish.

You may eat creamed cottage cheese which is usually the easiest to get.

After you've been on the Quick Weight Loss Diet for some time and taken off more than 30 pounds, you may add moderate portions of artificially sweetened jelly, plain yogurt, skim milk. However, you will lose faster if you don't add these foods.

As with all reducing diets, I suggest that you have no more than eight hours of sleep daily with this method. When you sleep more than eight hours, you're not assisting your body to burn up food most effectively and helpfully. You'll make reducing easier by keeping active and alert, on the go rather than on your back sleeping.

There's always a question in medical practice whether or not to tell a patient who is on a restricted diet what unusual sensations or symptoms may crop up in a very small percentage of cases. With a very impressionable person the doctor always runs the risk that such complaints may be introduced unknowingly even if they don't exist.

Another risk the doctor faces is that signs and symptoms may appear which are not imaginary but real and may be concealed by the patient because he is so anxious to reduce and doesn't want anything to slow him up. Therefore even though a high percentage of complaints are unfounded or imagined I must list those that do occur although very few individuals are so affected.

I've already mentioned dry, unpleasant taste in the mouth and on the breath which may occur if you don't drink at least eight glasses of water a day. Just drink enough water and this will not happen, also the diet will work most efficiently.

A very few people on the diet report some slight fatigue, a tired and sleepy feeling. This is readily corrected by a little quiet rest. Some very active people have said that they seem to lose some of their pep but gain it all back and more after two or three days. On the other hand, most people on the diet report a more buoyant, re-energized feeling almost from the start of the diet.

Experiments have shown that some individuals on a diet such as this, free of carbohydrates including sugars, do not manufacture the necessary sugar from the protein and fat. While the system in such persons does change protein to sugar (gluconeogenesis) and the body does use it, the result is not as effective as a sugar which comes from actual sugars, fruits, vegetables, and cereals. Accordingly a slight feeling of fatigue may develop. Such persons will respond immediately by taking a little sugar such as sweets or orange juice, and any fatigue will vanish. If the fatigue keeps recurring, switch to another of the diets listed later. (The feeling described should not be confused with the weak, trembly, perspiring symptoms due to a low blood sugar which a doctor's examination should reveal and which he would then treat accordingly.)

A very few persons on the diet say that the first day or two they've felt a slight light-headedness or giddy sensation on arising from lying down. This may be due to a drop in blood pressure or loss of sugar, water, or salt. This is rectified by arising slowly rather than rapidly from a horizontal or sitting position. I've found that such symptoms, if real, are very fleeting and invariably disappear after the first day or two on the diet.

It is vital to emphasize again that undesirable symptoms are *rare* among the many patients I've checked on this diet.

6. Combination Plan Keeps Weight Down: Quick Weight Loss And Stay Slim Eating

With the Quick Weight Loss method, followed as instructed, you will attain your desired or ideal weight. Now the problem arises how can I keep my weight down? How can I prevent that dangerous excess fat from creeping up on me again?

The combination of varied Stay Slim Eating and occasional return to Quick Loss dieting, if necessary, has made staying slim a happier, healthier way of living for thousands of women and men. It can do the same for you pleasantly and practicably.

Make no mistake about this – you do not go back to the eating habits that made you fat before. You must keep watching your weight. The happy surprise is that the techniques which have resulted in your losing weight make it easier to maintain ideal weight.

Instead of going on an eating binge after the diet, as some of my patients might have expected, they discovered that they had lost much of their exaggerated desire for rich food and overeating. You don't have as much trouble fighting temptation because that driving hunger has diminished.

Furthermore the majority of my dieters, and most hopefully you too, realize that as dependence on food and meals diminishes, your other pleasures increase. You usually become more aware that there is greater joy in other activities, in work and in play, in awareness of the world around you, in the physical pleasures of walking, looking, seeing. You're often released from such irritations and chronic ills as indigestion, lethargy, restricted movement, shortness of breath and other common symptoms of overweight. (Note well however that while weight loss is a great aid, it is not a cure-all. Any con-

tinuing aches, pains, or irregularities should mean a visit to
a doctor without delay lest you by-pass urgently needed
treatment for some ailment not due specifically to overweight.)
Not the least reward you get from this reducing is the surge of
pride in your proved willpower. These all-important assets,
plus your now habitually decreased appetite, help you form the
habit of effective Stay Slim Eating that keeps your weight
down. You also keep benefiting from the Quick Loss method
for the rest of your life by using it intermittently when
necessary.

Watch the 3-Pound Danger Signal

This is primary advice to put into effect to maintain desired
slimness. Get on an accurate bathroom scale unclothed the first
thing every morning as surely and regularly as you brush your
teeth before breakfast. 'The horridest of horror tales ... is
often told by bathroom scales.' Any time you see that scale
mark 3 pounds more than your desired weight, consider it
more serious than if your thermometer showed three degrees
or more over your normal temperature.

If you climb past that limiting 3-pound number, don't tell
yourself that 'I'll go easy next week,' or 'I'll take it off after a
while.' Take it off now. Starting that day or the next, go
back on the Quick Weight Loss Diet for a week.

Fundamentally your Stay Slim Eating from now on will
depend on calorie-counting as specified in this chapter. However, the big difference between this method and usual calorie-
counting on 'balanced dieting' as urged by others is the
combination with the Quick Loss method. This is of utmost
importance: *this combination is the crucial asset which turns
past failures into future rewarding success*. That vital variance
is why most of my patients have been able to maintain their
ideal weight year after year.

Case History

Specific examples of how well this combination dieting works are the cases of my collaborator, Samm Baker (5 ft 10½ ins., age fifty-six) and his attractive wife, Natalie (5 ft 4 ins., over forty).

A chubby youngster, Samm went off to college at sixteen and trimmed to his normal weight of 155. Over the years after college he varied between 155 and 165 by counting calories and limiting his intake accordingly. By fifty-four his weight had crept up to close to 170 and he varied between 162 and 170 although he wanted to get back to 155 for the sake of best health and appearance. He attributed his weight-losing difficulty to the demands of social and business eating.

Natalie had a serious weight problem as a youngster. Between thirteen and seventeen she weighed 167 pounds, wore size 18. Her fat was considered 'a family characteristic'. She hated being fat with all the social problems involved for a high school girl. Her family thought that fat was a mark of health and good living. Doctors gave her no support, said lightly that she'd 'grow out of it'. Summers at camp she would lose 10 or 15 pounds, then gain it all back after returning to mother's cooking and urging to 'eat, it's good for you'.

When she went off to college she determined to become slim. She invented her own rigorous diet, living primarily on soda water and peanut butter biscuits, skipping the dormitory meals. On vacations at home she had to pretend to eat because her mother was alarmed, then she'd return to her strict dieting at college. By age nineteen she weighed 109 pounds and wore size 8.

All through this two-year lopsided dieting she had never been ill even though she hadn't taken vitamins because the concept for everyday usage had not yet been developed. She had become beautiful, admired, and socially popular which is good indeed for a woman at any age.

Through the years of marriage and raising two children she maintained her weight by unceasing calorie-counting as she had 'a tendency to be heavy'. However, similar to her husband's case, her weight crept up as high as 130 at times (averaging about 125) although she felt best and believed she'd look best at 118. She couldn't get down to 118 on maintenance dieting.

At this point I introduced them to the Quick Weight Loss method for losing weight rapidly. They wrote out the diet, attached the sheet to the wall above their dinette table, and went on it together. Within a week each had lost 7 pounds. They returned to Stay Slim Eating for three weeks, then up went the Quick Loss Diet on the wall for another week. After the second week, Samm weighed 155 pounds, Natalie 118 pounds.

They've stayed at about this weight for over two years at this writing. Any time they gain over 3 pounds more than desired, they go back on the Quick Loss Diet for from three to seven days. They state that their desires for food have diminished. They both look and feel wonderful (as friends keep telling them), and they find enjoyable Stay Slim Eating 'a cinch' combined with the quick reducing diet.

Friends to whom they've introduced this method are just as enthusiastic, even more so when they're heavier to start with. They too find it easier to keep their lower poundage figures with an occasional return to the Quick Loss Diet.

7. How To Get The Stay Slim Eating Habit

From now on, Stay Slim Eating by calorie-counting – consciously or unconsciously – should become part of your way of daily living. It has been said that some people are terrible at counting calories, and have the figures to prove it! I disagree thoroughly with any doctors and nutritionists who state pessimistically that you can't get people to count calories. My patients have disproved this thousands of times when they follow the Combination Plan which so often makes the decisive difference between success and failure.

These Recommendations are Simple, Practical, Workable

1. Determine what your daily calorie intake must be according to the 'weight and daily calorie table' in Chapter 3. For example, if you're a woman 5 ft 4 ins tall, your total is 1,440–1,800 calories daily to maintain desired weight of under 120 pounds.

2. Start at once to get the habit of referring to the calorie tables on following pages to regulate your daily calorie intake.

3. Count the calories of everything you eat during the day, starting from your first food. In the beginning, write the figures on a slip of paper you carry or keep handy, and keep a running total. Stop eating for that day when you've reached the permitted amount. You'll find that soon you plan your day's eating automatically so that you apportion enough of your calories when you want them most, at dinner for example.

4. If you must exceed your total one day (because of a special dinner, for instance), deduct the excess from your next day's calorie total. Even if you must go on the fasting diet (Chapter

8) the next day to reach your two-day average, do it – you'll be helping, not harming yourself.

5. If the scale indicates the 3-pound overweight danger signal by the end of any week, go on the Quick Weight Loss Diet for the next week, or for three, four, or five days.

6. Then, after the Quick Loss interval, you start again with step one in this listing. You'll probably find to your intense gratification that the more often and longer you practise this combination, the easier it becomes for you to maintain your ideal weight.

Calorie Knowledge Becomes an Automatic Safety-Valve

As with most of my patients who have achieved ideal weight, you're likely to remember the general caloric content of foods after you've been counting calories only a week or two. If your remembrance is off a little now and then, it's not serious.

The tables that follow are your guide rather than a precise mathematical formula. No calorie table can be complete or exactly accurate because there is some variation in the size portions and individual foods. A 'medium orange' (65 calori is not something turned out to a specific size and weight o factory production line.

Your judgement must usually determine what is small medium, large. Your decision should err on the side of conservatism if you're doubtful. If you're not sure whether an orange, or a portion, is medium or large, call it 'large' and select the item or reduce the portion to what you can definitely classify as 'medium', or use a higher calorie count than for a medium portion. Keep in mind always that you're kidding yourself to your own detriment if you call a large portion a medium one. You may falsify a calorie total on paper, but your body adds correctly and will turn into fat any extra ories that you omit by fake mathematics.

I realize too that the size of portions as related to calorie count may be less clear than you (and I) would like them in the table. I've simplified and clarified the listings as much as I possibly could. You must contribute your own honest analysis and judgement with portions. 'One cup' of canned peaches, for example, refers to 8 ounces, and you must decide how many halves or slices that is. Because of variations in the size of the fruit, it just isn't possible to list 'six canned peach halves' rather than 'one cup'.

Similarly, the size of 4 ounces of steak can be determined by you either by weighing the steak or learning how big a piece makes a pound of steak and taking a portion one-fourth that size. You can judge 1 ounce by package content in many cases, as one-fourth of a 4-ounce package. (If listed in grammes, 1 ounce equals 30 grammes, 3⅓ ounces equal 100 grammes.)

Be sure you count everything you eat and drink as part of your daily total calorie intake. If you have a cocktail, for instance, that's fine – but remember that it must count in your mathematics as much as 150 calories of food. If you eat a lot of hors d'oeuvres at a dinner party, you must add the calories to your day's total.

As in Quick Loss dieting, you can appease your appetite with practically unlimited quantities of black coffee and tea with non-caloric sweetening, with hearty bouillon (see table, calories per cup), sugar-free jelly (10 calories per half-cup serving), and delicious non-caloric carbonated beverages.

By all means take advantage of the low-calorie foods available now to let you consume more calories in meat, vegetables, fruit, and other items that are particularly desirable for you personally. Keep in mind that if you have a 10-calorie sugar-free jelly instead of a 330-calorie slice of apple pie, that allows you 320 calories or an extra quarter-pound of steak and a baked potato.

Get into the excellent habit of drinking six or more glasses of water a day to flush out your system healthfully. If you feel very hungry at meals, make it a custom to have a glass of water, club soda, non-caloric beverage, or black coffee or tea before the meal. You'll find that you then have less desire to eat more than your daily calorie total permits.

Obviously it's simple to work out your own daily menus with this Stay Slim system of calorie-counting. Also use other dieting techniques detailed later, if desired. For example, suppose you're a woman 5 ft 4 ins and your daily allowance is 1,700 calories. Here's how you might combine calorie-counting and the Nibbler Diet (Chapter 8) for a day of satisfying eating:

Breakfast: 1 cup orange juice (110), black coffee – 110 cals

   10 AM: Black coffee or tea, 1 slice toast (60) with 1 tbsp jam (55) – 115 cals

   12 noon: Hamburger (minced beef) (245), 4 asparagus (15), half grapefruit (55), black coffee or tea – 315 cals

   4 PM: Black coffee, tea, or non-caloric soda, 1 slice toast (60) with 1 oz. cottage cheese (30) – 90 cals

   7 PM: Six shrimps with cocktail sauce (75), 2 lamb chops (280), baked potato (90), spinach (45), lemon meringue pie (300), black coffee or tea – 790 cals

   10 PM: 1 glass skim milk (90), 2 water biscuits (55) – 145 cals

With all this food you'd consume 1,565 calories and s... be 135 calories under your allowance of 1,700. You would... gain weight, especially by dividing the food into six fee... during the day. You could select these foods or any othe... that you prefer. Make up your own combinations to suit you... personal desires, as long as you stay within your Stay Slim... limit.

For instance, you could have a martini (150) before dinner and a large wedge of watermelon for dessert (120), eliminat... ing the pie (300).

Keep in mind combining calorie-counting with m... othe... suggested quick reducing diets for variety, taking adv...nta... of the features of each. For example, if you're tempted by... make your entire 300–350 calorie lunch in this case a slic...f angel cake (120) topped with a scoop of ice cream (200), ...ong with two cups of black coffee. You might have a cup of bouillon (10) first. You still won't gain weight and you needn't worry about any nutritional deficiencies. (This all applies to Stay Slim Eating, not while you're on the Quick Weight Loss Diet.)

Calorie Count Listing

Although calorie figures can't be entirely accurate, as noted, this listing will serve adequately as a guide for checking and totalling your calorie intake. While all foods can't possibly be included, these calorie counts will help you judge for most foods.

MEATS AND POULTRY

(Most meats and poultry figured here as lean, all visible fat trimmed off, about 60 calories per ounce.)

	cals
...on fried crisp, 2 slices	95
Beef, roast, lean, 4 oz.	210
Hamburger, lean, grilled, 4 oz.	245
Meat pie, 8 oz.	460
Steak, lean, grilled, ...oz.	235
...gna sausage, 4 in. medium slice, each slice	85

Chicken, turkey, grilled, 3 oz.	180
Drumstick and thigh with bone, fried, 5 oz.	275
Frankfurter, 1 medium size	155
Ham, smoked, 3 oz.	290
1 cup	320
Canned, all lean, 2 oz.	170
Lamb, chop, grilled, lean only, 2½ oz.	140
Leg, lean, 2½ oz.	130
Pork, roast, lean only, 2½ oz.	175
Sausage, 4 oz.	340
Tongue, ox, 3 oz.	205
Veal, cutlet, grilled, 3 oz.	185
Roast, lean, 3 oz.	280

FISH AND SHELLFISH

Clams, medium, each	9
Crabmeat, 3 oz.	90
Haddock, fried, 3 oz.	135
Herring, baked, 3 oz.	170
Mackerel, grilled, 3 oz.	200
Oysters, medium, each	12
Salmon, canned, 3 oz. drained	120
Sardines, canned, drained, 3 oz.	180
Shrimps, medium, each	10
Tuna, canned, drained, 3 oz.	170

VEGETABLES

Asparagus, medium, each	3½
Avocado, medium, half	185
Beans, baked and canned types, 1 cup	320
Beetroot, 1 cup	70
Broccoli, 1 cup	45
Brussels sprouts, 1 cup	60
Cabbage, raw, shredded, 1 cup	25
Cooked, 1 cup	45
Carrots, raw, 5½ ins., each	20
Cooked, 1 cup	45
Cauliflower, cooked, 1 cup	30
Celery, 8 ins. stalk, raw	5
Corn, cooked, 5 ins. ear	65
Canned, 1 cup	170
Cucumbers, 7½ ins., each	25
French beans, 1 cup	35
Lettuce, 5 ins. compact head, 1 lb.	70
2 large leaves	5
Lima beans, 1 cup	150
Mushrooms, 1 cup	30
Onions, raw, 2½ ins. each	50
Cooked, 1 cup	80
Parsley, raw, chopped, 1 tbsp	1
Peas, fresh, cooked, 1 cup	
Canned, frozen, drained, 1 cup	
Potatoes, medium, baked, with peel	105
without peel	90
Boiled, medium	90
French-fried, 2 ins. ½ in., each	15
Mashed, milk, butter, 1 cup	145
Chips, 2 ins. medium, each	
Radishes, raw, medium, each	3

Sauerkraut, drained, 1 cup	30	Grapefruit, 5 ins. medium, half	55
Spinach and other greens, 1 cup	45	Juice, fresh, 1 cup	95
Tomatoes, raw, medium	30	Grapes, 1 cup	85
Canned, 1 cup	45	Grape juice, bottled, 1 cup	165
Juice, 1 cup (8 oz.)	50	Lemons, medium, each	20
Vegetable Marrow, 1 cup	35	Oranges, medium, each	65
		Juice, fresh, 1 cup	110

FRUITS

Peaches, 2 ins. medium, each	35
Apples, raw, medium	70	Canned in syrup, pitted, 1 cup	200
Juice, 1 cup	125	Pears, 3 ins. medium, each	100
Apples, stewed, sweetened, 1 cup	185	Pineapple, fresh, diced, 1 cup	75
Apricots, raw, each	20	Canned in syrup, 1 cup	205
Canned in syrup, 1 cup	220	Plums, 2 ins. medium, each	30
⬛s, medium, ⬛h	85	Prunes, cooked, un- sweetened, each	17
⬛ries, black- ⬛rries, 1 cup	85	Juice, canned, 1 cup	170
⬛antaloupe, 5 ins. medium, half	40	Raisins, dried, 1 cup	460
Cherries, 1 cup	65	1 level tablespoon	30
Cranberry sauce, canned, 1 cup	550	Strawberries, fresh, 1 cup	70
⬛ates, pit⬛d, 1 cup	505	Tangerines, medium 2½ ins., each	40
⬛ig⬛ ⬛ried, large 2 ins. ⬛ 1 in. each	60	Watermelon, 4 ins.× 8 ins., wedge	120
⬛t Cocktail, ⬛canned, in syrup, ⬛ 1 cup	195		

DAIRY PRODUCTS, EGGS, FATS, OILS, DRESSINGS

Butter, 1 8-oz. cup	
(½ lb.)	1,605
1 pat or square	50
Cheese: Cheddar, 1-in. cube	70
Cottage cheese, creamed, 1 oz.	30
Cream cheese, 1 oz.	105
Process cheese, 1 oz.	105
Roquefort-type, 1 oz.	105
Swiss, 1 oz.	105
Eggs: large, cooked without fat, each	80
Scrambled, fried with butter, each	115
White only, raw, each	20
Yolk only, raw, each	60
Milk (cow's), whole, 1 cup (8 oz.)	165
Skim, nonfat, 1 cup	90
Buttermilk, cultured, 1 cup	90
Cream, single, 1 cup	525
,, 1 tbsp	35
,, double tbsp	50
Margarine, 1 cup (½ lb)	1,615
1 pat or square	50
Oils, cooking and salad – corn, olive, soybean, 1 tbsp	125

Salad dressings:	
French, 1 tbsp	110
Mayonnaise, 1 tbsp	110
Russian, 1 tbsp	75
Salad cream, 1 tbsp	60
Yogurt, plain, 1 cup	120

BREADS AND GRAIN PRODUCTS

Bread, all types, average slice, plain, toasted	60
Breakfast cereals, unsweetened, average, 1 oz.	110
Porridge, 1 cup	105
Rye crispbread, 2 ins. × 3½ ins., each	25
Saltines, 2 ins. square, each	
Macaroni, spaghetti, cooked, 1 cup	
Muffins, 3 ins. size, average, each	140
Noodles, egg, cooked, 1 cup	200
Pancakes, 4 ins., each	55
Rice, cooked, 1 cup	200
Rolls, medium size, average, each	130
Waffles, average size, each	
Water biscuits, medium, each	28

DESSERTS, SWEETS

Biscuits, sweet, average size, each	110
Cakes: Angel food, 2 ins. piece	110
Chocolate layer, 2 ins. piece	420
Cupcake, 2¾ ins., with icing, each	160
Plain cake, 3 ins. × 2 ins. × 1½ ins. piece	180
Sponge cake, 2 ins. piece	115
Chocolate syrup, 1 tbsp	20
Doughnuts, medium, plain, each	135
Gelatin dessert, ½ cup	80
sugar-free	10
Honey, 1 tbsp	60
Ice-cream, ½ cup	200
Ice-cream soda, average size	350
Jelly, ½ cup	80
sugar-free	10
Jams, preserves, 1 tbsp	55
Pies, apple, other fruits, 4 ins. piece	330
Custard, 4 ins. piece	265
Lemon meringue, 4 ins. piece	300
Puddings, custard, 1 cup	275
Sherbet, ices, ½ cup	120
Sugar, granulated, 1 tsp	16
Sugar granulated, 1 cup	770
Sweets, caramels, fudge, 1 oz.	120
Chocolate, milk or plain, 1 oz.	145
Syrup, 1 tbsp	55

MISCELLANEOUS

Beverages: Coffee, tea, plain	0
Beer, 8 oz.	110
Cocktail, average	150
Gin, Scotch, vodka, average, 1 oz.	75
Carbonated beverages, ginger ale, 8 oz.	80
Cola-type, 8 oz.	105
Cocoa, cup	235
Ketchup, chili sauce, 1 tbsp	15
Olives, green and ripe, large, each, average	9
Nuts: Peanuts, roasted, shelled, ½ cup	420
Peanut butter, 1 tbsp	90
Cashews, pecans, walnuts, ½ cup	375
Pickles, dill 4 ins., sweet 3 ins., each	18

Pizza, cheese, 6 ins. wedge	200	Chicken, tomato, vegetable	80
Soups: Bouillon, broth, consomme, 1 cup	10	Creamed, asparagus, mushroom, 1 cup	200
		Rice, noodle, barley, 1 cup	115

Make a Game of Maintenance Eating

Your positive attitude is one of the most important elements in maintaining your slim figure once you attain it through Quick Weight Loss dieting. Instead of grumbling about calorie-counting, make a game of it, just as many people enjoy working out number games and puzzles. Your game challenge is to pattern the most desirable eating you can figure out each day within your personal calorie allowance. There are dozens of combinations you can work out, so enjoy arriving at the answer that will please you most each day. After a week or two it will become almost second nature to you to know how much you can eat of what foods.

Never overlook or delude yourself about 'hidden' calories. For example, if you add butter to your vegetables, you must add in the butter calories to your total daily intake. It becomes very easy to cut out such extra, self-pampering calories. I have had patients who said they couldn't possibly even think of eating a baked potato or vegetable unless drenched with butter. When they did give up the butter because of the weight-producing calories on the Quick Weight Loss Diet, they told me eventually that now they couldn't stand the thought of 'spoiling the delicious natural flavour' of the good food with rich butter. This will undoubtedly happen with you if you give it a reasonable trial.

8. More Quick Reducing Diets For Variation

Note this again for your quick slimming success: for speedy reducing and staying at ideal weight, in combination with Stay Slim Eating the Quick Weight Loss Diet has been used more effectively by more of my patients and others to whom I've recommended it than any other reducing diet.

A very small percentage of overweights in my practice wouldn't stay with the Quick Weight Loss Diet for one reason or another. Some didn't want to do without fruits and vegetables even for a week or two. Some had an aversion to drinking eight glasses of water a day. Others couldn't explain why the diet was not right for them.

Since my goal has always been to reduce the overweight individual – to *get that fat off* – rather than to force a particular diet, I worked with each person to devise the diet that would take off weight most effectively and surely in each case. This has resulted in my compiling a variety of quick-reducing diets, some perhaps bizarre (but effective) listed here with my personal specifications.

I recommend the same to you, to shift your reducing methods if necessary from one to another listed here. If one quick reducing diet is very difficult for you to follow, or seems to bring on undesirable symptoms – whether real or imaginary – choose another more to your liking from the many in this chapter.

The one all-important point is to get that fat off you before it impairs your health drastically and irreparably. Every extra ounce is an extra threat according to proved, inescapable statistics. The more excess weight you carry, the less your chance of enjoying maximum vigour, health, and attractiveness and living out your full years.

It's unfortunate that many dietary advisers overlook the vital fact that persons are as different and individual as fingerprints. With that in mind I've worked out for you here the many diets that I've created over the years for my patients. I believe that among these you will find the one diet or the right assortment of diets, that will take off weight rapidly and then keep it off for you as an individual. What has happened successfully for so many others can and should work for you.

The Quick Loss method and all the following quick reducing diets have reduced individuals where 'balanced' dieting methods have failed. Some people have found it helpful to switch from one of the diets to another after a week or two, for variety. The choice is yours. Either the Quick Loss or the other diets here can be your most productive route to a slimmer, trimmer figure and a new dimension in vigorous, healthful living.

Low-Fat or 'Fat-Free' Diet

This low-fat or 'fat-free' diet has one fundamental rule: fat in any form, visible or invisible, is eliminated. You have a wide choice of foods but you must not devi...

...this diet for mon... ...at your... Afte...

'YES' FOODS, YOUR CHOICE:

'NO' FOODS, DO NOT EAT:

Meats: grilled, boiled, or roasted beef, beef liver, lamb, veal, chicken, turkey – all with all visible fat removed

NO bacon, bologna sausage, frankfurters, ham, pork, salami, sausage. No smoked meats. No smoked poultry. No duck

Fish and Shellfish: carp, cod, flounder, haddock, or white perch. Clams, crabmeat, lobster, mussels, oysters, scallops, shrimps

NO salmon, sardines, tuna fish. No smoked or fried fish or shellfish

Vegetables canned, fresh, or frozen: asparagus, eggplant, green beans, lettuce, peas, potatoes (boiled, baked, mashed, no butter or margarine), spinach, and all greens, all types of marrow, French beans, tomatoes. (Gaseous ve doctor

need not stay with just the foods listed in the preceding which is a sample selection. Choose the foods for your six meals to your own liking according to calorie counts in the calorie table, for a daily total of up to 1,200 calories.

Some days you may find that you have consumed from 1,250 to 1,500 calories. Don't be disturbed about occasional lapses. Cut back for a few days to about 900 calories, then maintain 900 to 1,200 calories daily. My women patients on the nibbler diet reduced with even 1,200 to 1,500 calories daily, and men on 1,200 to 1,800 calories. For desirable quick weight loss, start with 900 to 1,200 calories daily and then you may increase after you've learned control.

On the six-meals-a-day nibbler diet, you will lose from 3 to 6 pounds a week depending on your amount of excess weight and size, and whether you hold your daily calorie count to 900 or allow yourself more. This is a successful form of healthful dieting that you can stay on for the rest of your life.

It's important that you don't fool yourself by eating one or more heavy meals and then having the three extra meals also. You must stay within the daily calorie totals permitted in order to lose weight.

Weigh yourself every morning. If, due to circumstances such as extensive entertaining, you have gained 3 pounds or more, go on a fasting diet for a day or two, or the Quick Weight Loss Diet for a few days to a week, or on one of the One-Dimensional diets. When you've taken off the excess 3 pounds or more, go back to the nibbler diet and stay with it. If you keep slipping off the diet, then switch to the Quick Loss Diet or another of those in this book to find the one that works best for you.

Unless you're one of the 5 per cent who have some metabolic disturbance, you cannot fail to lose weight on the nibbler diet.

Important: the nibbler dieting technique of spreading the daily consumption of food over six meals instead of three meals may be applied effectively to any of the diets in this book.

You will reduce more efficiently and swiftly by the six-meal method no matter which diet you choose.

Low-Salt Diets

A low-salt diet reduces you in two ways. First, the selection provided is primarily from low-calorie foods. Second, the foods are so bland and there appears to be such a sameness in many foods that the dieter unconsciously cuts down on quantity and consumes less than 1,200 calories daily. Some may over-indulge for the first few meals and then find that desire decreases. Because of the blandness of the foods, the appetite is curbed habitually. Most of my patients on this diet have lost weight steadily. They had no difficulty in maintaining ideal weight when they reached it. Some have lost 50 pounds or more, and then stayed slim.

Here is the listing for the low-salt diet that has proved effective for many individuals:

General rules: don't deviate from the listing that follows, as with every diet here you must stay completely within the restrictions. Eat only enough at each meal to satisfy your hunger, then stop. No salt on the table; salt substitutes may be used. No salt used in cooking. No fried foods. No salted butter. Only salt-free bread. Only fresh vegetables, meat, and fish – no canned or frozen.

'YES' FOODS, YOUR CHOICE:	'NO' FOODS, DO NOT EAT:
Meats: beef, lamb, veal; chicken, turkey. Brains, heart, kidney, liver	NO salted, smoked, canned, or pickled meat, no pork or bacon, no canned meat gravies
Fish: Fresh cod, flounder, haddock, halibut, salmon	NO shellfish, no salted, smoked, canned, or pickled fish, no canned fish gravies

'YES' FOODS, YOUR CHOICE

'NO' FOODS, DO NOT EAT:

Vegetables: artichokes, asparagus, avocados, beetroots, carrots, eggplant, green beans, lettuce, fresh peas, potatoes, vegetable marrow (all types), spinach, and all greens, tomatoes. Also, broccoli, brussels sprouts, cabbage, cauliflower, corn, cucumbers, onions, peppers, radishes, turnips

NO canned or frozen vegetables or vegetable juices, no beans, celery, lentils, or dried peas

Fruits: all fresh fruits and canned, cooked, and frozen fruits, all fresh, canned, and frozen fruit juices, no-salt fruit sauces (apple, cranberry, etc)

Dairy Products: Unsalted butter, sweet or sour cream, cream cheese, cottage cheese, milk, dry, or evaporated skim milk

NO salt butter, no cheeses unless named in 'yes' listing

Breads and Cereals: salt-free only! Whole wheat and enriched cereals and bread, salt-free cereals, salt-free rice, macaroni, noodles, spaghetti

NO breads or cereals that are not definitely salt-free

Desserts and Miscellaneous: Jelly, tapioca, olive and vegetable oils, unsalted nuts, honey, cocoa, coffee, tea. Flavourings: cinnamon,

NO commercial cakes, biscuits, puddings except definitely salt-free, no commercial mayonnaise or salad dressings, no peanut butter, no

'YES' FOODS, YOUR CHOICE:	'NO' FOODS, DO NOT EAT:
cloves, garlic, green peppers, lemon juice, mace, mushrooms, nutmeg, onion, paprika, vinegar	salted nuts, no commercial ice cream, ices, or sherbets. *No ketchup*, no mustard, spices, or sauces unless in 'yes' listing

Here's a sample day's menu, using moderate portions:

Breakfast	Lunch	Dinner
Fruit or fruit juice	4 oz. meat, fish or cheese	4 oz. meat, fish or cheese
Any one: cereal or egg or slice of bread with unsalted butter	Any one: 1 slice bread with unsalted butter or rice or potatoes or noodles	Any one: 1 slice bread with unsalted butter or rice potatoes or macaroni
Milk, coffee, or tea	Vegetable or salad	Vegetable or salad
	Fruit or plain dessert	Fruit or plain dessert
	Milk, coffee, or tea	Milk, coffee, or tea

It's best to check portions of what you eat on the calorie table, and consume about 1,200 calories daily, as in the preceding sample day's diet. You may not lose much the first few days but then you will average weight loss of 2 to 6 pounds a week, depending on how much you're overweight. You can stay on this diet indefinitely but unless you should remain on low-salt intake for personal medical reasons you can go on to normal Stay Slim Eating once you're at your desired or ideal weight.

Low-Salt Low-Cholesterol Diet

This excellent variation of the low-salt diet also reduces the cholesterol intake for those who should avoid cholesterol. Follow the preceding Low Salt diet but eliminate the following foods: no whole milk, no cream, no sweet or salt butter, no egg yolk. Lean meat and poultry only, all visible fat trimmed off, lean fish only. No cocoa. All the other foods in the Low-Fat 'no' list should not be eaten.

One-Dimensional Quick-Loss Bizarre Diets

While most of my patients have succeeded in losing weight quickly and happily on my primary recommendation, the Quick Loss Diet, a small percentage have found that a one-dimensional diet was the best to get weight off in a hurry for them. On the bizarre one-dimensional diet where they were restricted to one food only, they were able to exercise their willpower best, give up all thought of temptation with any other food whatsoever, and lose weight rapidly.

Here are some of the bizarre diets which have proved effective. Some general rules must be followed. A therapeutic vitamin-mineral tablet should be taken daily as a safety measure even though it may not be necessary with some of the diets. Some of my patients have been on such bizarre diets for two weeks to two months without ill effects, but you should stop if you feel fatigued, lightheaded, or disturbed. If you should feel faint, a little orange juice or a toffee should be taken. After eating more for a day or two, you might try again if you're determined. No one with a health disorder should undertake these bizarre diets.

On the other hand, many patients who could not reduce any other way have found that they lost weight quickly and felt great after going on one or more of these bizarre diets. Some even followed one of the diets for a week, then went on another of the One-Dimensional diets, back to Stay Slim Eating, and have continued intermittently in this fashion.

Like some of my patients, you may read one of the bizarre diets, like it instantly and exclaim, 'That's for me!'

On all the bizarre diets you may have as much as desired of coffee and tea without cream, milk, or sugar, but with non-caloric sweeteners. You may take unlimited quantities of non-caloric carbonated beverages and calorie-free bouillon and consommé. Drink at least eight 10-ounce glasses of water daily.

On the average, those on the bizarre one-dimensional diets lose from 5 to 15 pounds a week, depending on amount of overweight and the size of the individual. The bizarre diets have proved safe and highly effective for many women and men within my experience.

I've had patients eat one hard-boiled egg daily for two weeks; 8 ounces of cottage cheese daily for sixteen weeks; one chicken leg daily for two weeks; one 4-ounce hamburger daily for four weeks; one portion of grilled fish every other day for one month. In such cases it's absolutely essential to drink eight to ten glasses of water daily. For these individuals, with no serious illness or recent operation, no harm was done. Because of the drastic change of eating habits, a slight fatigue, headache, or light-headedness may develop. Such symptoms usually pass as the body adjusts itself. If they continue, a doctor should be consulted.

Buttermilk-Only Diet

This is a very simple diet, as are most of the One-Dimensional diets. Six glasses of buttermilk daily, spaced out one glass every three hours. The only other foods permitted are coffee and tea without cream, milk, or sugar, but with non-caloric sweeteners, also as much as you want of non-caloric sodas and calorie-free bouillon and consommé, along with a vitamin tablet daily.

Cottage Cheese and Grapefruit Diet

Moderate portions of cottage cheese and grapefruit – six times daily. Drink as much as you wish of artificially sweetened carbonated beverages, and coffee and tea with artificial sweetening, no cream or milk. If you prefer, you may substitute melon for the grapefruit, or alternate the fruits, using melon one day and grapefruit the next.

Cottage Cheese and Buttermilk Diet

For those heavies who are very fond of cottage cheese and buttermilk, restricting their eating to these two fine foods has proved effective and healthful in taking off 10 pounds and more in a hurry.

The diet consists merely of eating as much cottage cheese as wanted, and a glass of buttermilk, up to six times daily. This rarely adds up to more than 540 to 710 calories daily. Patients on this diet say that they find the foods filling and satisfying and they have no hunger pangs. When taken with vitamin tablets and large quantities of water and permitted non-caloric beverages, no deficiencies have developed in patients even over a period of months on this diet.

As with all one-dimensional diets, if you have an aversion to the principal foods, don't even begin! If you're going on a one-dimensional diet, choose the particular one that appeals personally to your taste.

Baked Potato and Buttermilk Diet

One baked potato daily, complete with skin, six glasses of buttermilk spaced out during the day. I've known some very heavy individuals to stay on this diet for months. As should be done in every such case, the blood and urine were checked particularly for any deficiencies. In practically every case I found the individuals in sound physical condition with health much improved by the sizeable losses in weight.

Eggs and Tomatoes Diet

This diet is a favourite among some of my veteran reducers who wanted to get off weight rapidly at the start. They continue to use this as a quick-action diet to take off quickly pounds which may have crept up on them. *They simply eat a hard-boiled egg and a small tomato up to six times a day, and only when they feel hungry.* The fat vanishes quickly day after day.

This and similar bizarre diets work amazingly for many heavy eaters because they undertake the one-dimensional diet as a challenge for a limited period. Thus they're able to wipe away any desire to eat just for the sheer enjoyment of packing in tasty food even when not hungry. Being acutely conscious of what they've set out to accomplish on the bizarre diet, they don't eat out of boredom or just for the pleasure of eating. I've seen some of the worst gluttons go on this Eggs and Tomatoes diet, for instance, and stay with it. They lost dangerous and burdensome fat that they couldn't get off before.

Meat-Only Diet

This diet is limited to a half-pound of any meat wanted, three times a day, along with plenty of water and the other non-caloric drinks. For lots of those who like meat very much, this has proved an effective quick-reducing diet. A doctor I know of has used this as the prime reducing diet for many of his obese patients and those ill from excess weight. He claims that he finds this diet so successful and safe because 'meat and water are the only two natural foods on this planet.' One might question his statement but I can attest that a good many over-weights drop excess fat rapidly with this meat-only diet.

On this diet, eat lean meat, only when hungry and only

enough to take away hunger. You'll lose great amounts of fat quickly, safely, and surely.

Poultry-Only Diet

Follow the same instructions as in the preceding meat-only diet but instead of meat, use chicken or turkey with the skin removed.

Fish-Only Diet

This is another variation of the meat-only diet; it has been used successfully by those who are very fond of fish and shellfish. It affords a large choice from a wide variety in those categories. Simply follow the same instructions as the meat-only diet but choose from any shellfish and lean fish.

Meat-, Poultry-, Fish-Only Diet

This diet affords you a combination of meat, poultry, fish, and shellfish during any one meal or confining one type to each meal. For example you might have grilled fish at breakfast, chicken at lunch, and lean meat at dinner. You might enjoy at dinner some shrimps, a small chicken leg, and some lean meat. In general follow the instructions for the meat-only diet.

All-Vegetable Diet

Eat vegetables only, as much as you want of vegetables of your choice, up to six meals daily. This means exactly what it states – *vegetables only* – no nuts, butter, milk, eggs, or cheese. Drink a good deal of water and unlimited quantities of the noncaloric liquids.

For many individuals, this can be a lifetime diet. If you're sold on it, can tolerate it, if no deficiencies show up in regular medical check-ups, then you can feel assured that you'll never be heavy again if you become an all-vegetable eater.

You may eat practically any vegetables except a few such as avocados, beans, and lentils, choosing from this list.

	CALORIES		CALORIES
Asparagus, 1 cup	70	French beans, 1 cup	35
(8 oz.)	35	Horseradish, 1 tbsp	25
Beetroots, 1 cup	70	Lettuce, hard head	70
Broccoli, 1 cup	45	Mushrooms, 1 cup	30
Brussels sprouts, 1 cup	60	Onions, 1 cup cooked	80
Cabbage, 1 cup		Onions, 1 cup raw	50
cooked	40	Parsley, 1 tbsp	1
raw	25	Potato, 1 medium baked	105
Carrots, 1 cup	45	Radishes, 1 cup	25
Cauliflower, 1 cup		Sauerkraut, 1 cup	30
cooked	30	Tomato, 1 medium	
Celery, 1 cup	20	fresh	30
Corn, 1 small ear	65	juice, 1 cup	50
Cucumber, 1 cup	20	Turnips, 1 cup	40

If you ate most of the portions listed above in one day, totalling 16 cups of vegetables, plus 1 tomato, 1 good-sized head of lettuce and a medium-sized potato, you'd still be eating only about 900 calories. It's not likely you could eat all this in one day, so you can readily see that your calorie intake is small on an all-vegetable diet.

You can make the foods more interesting with spices and a sprinkling of low-calorie dressing. You may vary the vegetables with fruits of comparable caloric value, referring to the calorie charts. If you stick to it, you'll take off weight steadily and you'll never be fat again.

The big danger on an all-vegetable diet is if you start to relax and add vegetable oil dressings, bread, butter, cream cheese, eggs, nuts, milk and such. Before you know it, you add 500 to 1,000 extra calories per meal and up goes your weight. So, as with all restrictive diets, remember that there must be no deviations and no extras.

Fruit-Only Diet

If you love fruit, this diet may prove to be exactly what you want for rapid weight loss over the period of a week or two, as it has for others. Here's a sample day's diet:

Breakfast: 1 orange whole or sliced, or 2 prunes, or ½ grapefruit

Lunch: 1 small melon, or 1 large slice of a large melon plus 1 large slice of watermelon

4 PM: 1 peach (or equivalent available fruit)

Dinner: 1 pear, 1 orange, 1 apple, 1 plum

10 PM: 1 apple

At each meal you may have coffee and tea without milk, cream, or sugar. Throughout the day, drink as much as you wish of non-caloric carbonated beverages.

Your total intake on this diet will be only about 350 calories per day. If you really like fruit you'll enjoy the eating as you lose weight rapidly. Remember that this is a fruit only diet and therefore nothing else is permitted, no lettuce or salad dressing or anything else not listed in the preceding. You may substitute other available fresh fruits for those listed.

Vegetable and Fruit Mix Diet

This popular diet for those who like both fruit and vegetables is simply a combination of the two. Refer to the previously listed vegetable diet and fruit diet, mix the two, consulting the calorie table, and being certain that your daily intake doesn't exceed 900 calories. You'll find that you'll have to eat a lot to reach that calorie total. As with all these diets, eat only to satisfy your immediate hunger, do not stuff yourself beyond that point.

You may spread your vegetable and fruit consumption over six meals a day, as described in the Nibble diet. Again, have all you want of coffee and tea without milk, cream, or sugar, and of non-caloric carbonated beverages.

Quick Loss Fruit and Vegetable Alternating Diet

This series of alternating quick loss diets works extremely well in reducing many patients who are considerably overweight and others who want to lose 10 to 20 pounds:

First week: Quick Weight Loss Diet
Second week: Fruit Only Diet
Third week: Quick Weight Loss Diet
Fourth week: All Vegetable Diet
Fifth week: Quick Weight Loss Diet
Sixth week: Vegetable and Fruit Mix Diet
Seventh week: Quick Weight Loss Diet
Eighth week: Fruit Only Diet

Continue rotation or alternate the weekly diets any way that suits you best. This system satisfies the desire of many for differing tastes so that each week's eating is looked forward to with anticipation. The change from no fruit and vegetables on the Quick Weight Loss Diet to an abundance of both is enjoyed, and the excess pounds keep coming off surely, swiftly, steadily.

Those who are not very overweight usually can shift to Stay Slim Eating after a few weeks. This specific alternating system can be adapted with many of the dozens of diets listed here. The choice is yours to attain that one essential goal: taking off weight rapidly and keeping it off for the balance of your lifetime.

Smaller Portions Minus Diet

This simple diet depends on cutting down to smaller portions of the foods you eat normally, *minus* certain high-calorie items. Each portion for needed losses should be about one-third of the usual amount. This diet is easy to calculate but it requires a strong will to resist larger portions of the foods you enjoy and are accustomed to eating in greater quantity. A typical day's eating would work out as follows:

USUAL PORTIONS		SMALLER PORTIONS MINUS	
Breakfast:	cals		
8 oz. orange juice	110	3 oz. orange juice	40
2 fried eggs in butter	230	1 boiled egg	80
2 slices buttered toast	220	1 slice toast, no butter	60
coffee with sugar, cream	65	black coffee	0
Lunch:			
6 oz. hamburger	360	2 oz. hamburger	120
buttered roll	180	no roll	
8 oz. peas with butter	160	3 oz. peas, no butter	40
3 biscuits	330	1 biscuit	110
coffee with sugar, cream	65	black coffee	0
Dinner:			
martini	150	1 oz. whisky	75
8 oz. chicken soup	80	8 oz. consomme	10
8 oz. steak with butter	520	3 oz. steak, no butter	180
baked potato with butter	205	baked potato, no butter	105
8 oz. spinach with butter	95	8 oz. spinach, no butter	45
apple pie	330	half grapefruit	55
coffee with sugar, cream	65	black coffee	0
Total: 3,165			902

While most of the portions have been cut about one-third, some substitutions have been made, and 'minus' items include butter, cream, sugar, pie. It may astonish you that the smaller portions minus method has eliminated 2,245 calories for the day. If the eater here were a man 5 ft 10 ins tall, his calorie intake for ideal weight would be about 2,000. On usual portions eating he would have consumed an excess of 1,165 calories to add to his fat. Or, with smaller-portions-minus, he'd eat 1,080 fewer calories than maintenance, helping to reduce his weight.

On a maintenance diet he could increase the size of his smaller portions to total about 2,000 calories and stay slim. If this kind of dieting appeals to you, and you have the will-power to keep from reaching for larger and extra portions, it's clear that you can cut down weight steadily this way.

You must keep in mind this difference from the usual suggestions of others merely to eat smaller portions in order to reduce: on my smaller portions diet you cut out completely those high calorie foods specified. I have found that this usually makes the difference between success and failure in this type of dieting.

Bananas and Milk Diet

If bananas are a favourite food of yours, and you like skim milk, this is a pretty well-rounded diet that will take off weight quite rapidly. A banana is 75–100 calories and a glass of skim milk is 90 calories. By considering a banana as one meal, and a glass of skim milk as another meal, and alternating them, you can eat nine spaced-out 'meals' a day and still total only 900 calories.

This is generally a filling, satisfying diet, involving no hunger pangs. You get plenty of protein as the milk contains 10 grammes of protein per glass; each banana has 1 gramme of protein also.

Juices-Only Diet

This juices-only diet is a favourite among a good many Europeans for taking weight off in a week or so. The diet consists only of natural vegetable and fruit juices: eight 10-ounce glasses spaced out during the waking hours. No sugar is permitted, but it's fine to sweeten with non-caloric sweeteners. Remember, at least eight glasses of water are also to be taken daily, plus as much as wanted of coffee and tea without milk, cream, or sugar, and non-caloric carbonated beverages. Take a vitamin pill daily. Weight comes off rapidly. This is a good two-week reducer, but may be extended for months without fear of protein deficiency.

Suit Yourself One-Dimensional Diets

Some of my patients concoct their own one-dimensional bizarre diets and have been very successful at taking off weight fast in a short time. To suit yourself, choose one or two favourite foods that are naturally large in bulk and low in calories. Figure out the quantity that will total less than 900 calories a day. Then space out the number of meals you can make from this quantity, supplementing as usual with plenty of water and the non-caloric liquids.

As one instance of a personal selection, a famous judge I know went on a one-dimensional watermelon diet during the summer. He lost 50 pounds and then went on Stay Slim Eating the rest of the year, never gaining back the 50 pounds during all those months.

Yogurt-Only Diet

Quite a few patients have lost weight quickly and pleasantly on a yogurt-only diet. For this diet take 120 calories of yogurt five to six times daily (total of 600-720 calories). You may vary the flavours as much as you please, keeping the total calories under

900 daily. Drink plenty of water and additional non-caloric liquids. If constipated, take milk of magnesia or enemas.

Repetitious Same-Meal Diet

Patients on this diet have told me that they follow it easily because all they have to remember is the food for one meal. They eat that same meal for breakfast, lunch, and dinner, as follows:

> 1 slice of toast with dietetic jam
> 4 oz. fruit or vegetable juice
> 3 tbsps cottage cheese
> coffee or tea without milk, cream, or sugar, as much as
> wanted, also non-caloric carbonated beverages

Each meal is about 170 calories. You may eat five meals spaced out (such as 8 AM, 12 noon, 3 PM 6 PM, 10 PM) for a daily total of 850 calories, or three meals for a total of only 510 calories. Your weight loss will be quicker on the three-meal routine, of course, but even on 850 calories you should lose rapidly.

If you find cottage cheese more filling and satisfying than the juice, you may eliminate the juice and add 3 tablespoons of cottage cheese (total 6 tablespoons per meal); you'll consume the same low number of calories, resulting in rapid weight loss.

Skipper's One-Meal-a-Day Diet

Are you like the woman described in the following? She said, 'I found I was gaining weight rapidly because no matter how much I ate during the day I had to eat a big meal with my hus-band at night. I went to my doctor who put me on a "balanced diet" of a "balanced" breakfast, lunch, and dinner. But it didn't work for me. My husband insists on a good dinner at home or in a restaurant. I ate along with him and kept putting on weight. What can I do?'

It turned out that this woman didn't care for breakfast. At lunch she was at work and that meal wasn't important to her either. But she ate morning and noon because she was afraid to forsake 'balanced eating'. Her solution was simple: when she was convinced that she wouldn't be harmed by drinking black coffee for breakfast, then black coffee with half a grapefruit, an orange, apple, or pear at lunch, then a good dinner, she started losing weight.

Because she became accustomed to eating little during the day, she found that she was contented with less at dinner too. After six months she was at her ideal weight and went back to the same doctor for a check-up, telling him how she was eating. He examined her, then said, 'You're eating "crazy", but your blood pressure, blood count, electrocardiogram, are just wonderul. Keep doing what you're doing, it's good for you!'

I've checked many such patients with the same fine results. If you're an individual who enjoys dinner but doesn't care for breakfast or lunch, this One-Meal-A-Day skipper's diet may be just right to take off your excess weight easily and pleasantly. Here's the skipper's diet that has worked well for many:

Breakfast: coffee or tea, 4 oz. skim milk if desired, and non-caloric sweetening, also a small glass of tomato or orange juice if wanted

Lunch: fruit (half grapefruit or an orange or apple) and coffee or tea with milk and non-caloric sweetening if desired

Dinner: ½ lb. lean meat (or caloric equivalent in poultry or fish – see calorie table)
salad with low calorie dressing
half grapefruit or melon or 1 fresh fruit
black coffee or tea

2 hours later: (if desired) coffee or tea and a piece of fresh fruit; or a glass of skim milk with a slice of toast and 1 tbsp cottage cheese

2 *hours later*: (if desired) skim milk or yogurt or a piece of
fresh fruit; or coffee or tea with a slice of thin toast and a
thin slice of Cheddar cheese

The dinner totals 600 to 700 calories and the day's eating
1,000 to 1,100 calories. Continue it – varying only the meat,
poultry, or fish – until you're at desired weight. Then increase
to Stay Slim Eating as previously detailed.

Skipping meals is not recommended for those who tend to
become fatigued, lightheaded, or trembly. To find out whether
such symptoms are real or imaginary, have your doctor take a
blood sugar test at about 10 or 11 AM. If your blood sugar is
normal, the symptoms bear no relationship to lack of energy in
your blood.

People who are free of any such symptoms can go on the
skipper's diet for as long as necessary to reduce to normal. Once
reduced, ladies, don't become so delighted with your new slimness that you continue to lose weight. This is not dangerous
but is unnecessary and may produce listlessness. If you're a
standard size 10 or 12, don't try to prove to your friends that
you can get into a 9 or 7 dress, as some women do. This form
of reaction is not found in men; they're usually content to
become slim but not scrawny.

Beware on this diet of packing away a tremendous meal at
night. This is especially dangerous to the arteries, which
may become choked with fat, and taxing to the pancreas,
sometimes leading to the production of diabetes.

To help curb any desire to eat too much at dinner, it helps
greatly to fill yourself with liquid first. Before sitting down to
dinner drink one to three glasses of water, or coffee or tea without milk, cream, or sugar, or non-caloric beverages. You may
not find this necessary after you're used to the skipper's routine.
The excellent meal listed will slim you down and satisfy you
too.

Liquid and Other Formula Diets

The liquid and other formula diets packed in cans and other packages have been helpful to many people and may be effective for you in losing 2 to 5 pounds a week. Even the general medical profession has accepted the fact that they are not harmful to the average overweight, but have benefits for many.

It helps keep intelligent perspective to realize that formula diets are not new. Infants have been on formula diets for many years, or breast fed which is in effect a formula diet. The Rockefeller Institute had some individuals on liquid formula diets for as long as four years. Many doctors made up their own powdered foods for patients to mix with water or milk.

When the commercial formula dieting products became so popular a few years ago the first outcry from many doctors and nutritionists was *against*. They protested that this was drastic dieting at its worst! Then a more reasonable viewpoint took over. Dr Vincent P. Dole stated in an editorial in the NY State Medical Journal that, '... formula diets are not radically new. Physicians have used liquid diets in the treatment of disease since at least the time of Hippocrates, and more recently have employed nutrient mixtures for complete feeding of infants.'

While emphasizing that 'liquid mixtures do not have magic properties', Dr Dole agreed that liquid dieting is not dangerous for it has the same nutrient qualities as any nutritive solid diet, and has the added factor of exact calculations not found in many low calorie diets.

If the liquid formula diet appeals to you individually more than the Quick Loss or other diets here, go on it with specific understanding. One advantage of the packaged diets is that if you wish to add 225 more calories a day, you simply take another portion. However, this is a highly disciplined, monotonous way of dieting. Be prepared to stay with it for at least three months to take off a lot of weight. Go at it gradually, not all-liquid at once, as I advise in the following liquid formula diet. If not, you may develop a slight diarrhoea or a bloated feeling of discomfort which may cause you to drop the diet. The following will help you succeed, on a formula diet.

Liquid Formula Diet

First 3 days: 1 glass for breakfast, then 675 calories of varied regular foods divided into lunch, dinner and/or snacks (use calorie table to figure out calories). For example, choose from lean meat, poultry, or fish, salad with low-calorie dressing, fresh fruit, low-calorie vegetables etc

4th to 6th days: 1 glass for lunch, and 675 calories of varied foods for other meals

7th to 9th days: 1 glass for dinner, and 675 calories of varied foods for other meals

10th day: 1 glass for breakfast, 1 glass for dinner, 450 calories of varied foods for lunch

11th day: 1 glass for breakfast, 1 glass for dinner, 450 calories of varied foods for lunch

12th day: 1 glass for breakfast, 1 glass for lunch, 1 glass for snack, 225 calories of varied foods for dinner

13th day: 1 glass for lunch, 1 glass for dinner, 1 glass for snack, 225 calories of varied food for breakfast

14th day: 1 glass for breakfast, 1 glass for dinner, 1 glass for snack, 225 calories of varied foods for lunch

Thereafter: all-liquid formula diet, 4 glasses a day for breakfast, lunch, dinner, snack – total 900 calories

You may interchange 225-calorie formula portions of wafers or soups with the drink if you wish, at any time.

In addition to the formula, drink as much as you want of water (the more the better), coffee and tea without milk, cream, or sugar, and non-calorie carbonated beverages.

Formula and Alternate Diets

Some of my patients, because of their individual characteristics which may be like yours, have reduced very happily by alternating a liquid formula diet with other quick-action diets listed in this chapter. For example, you can go on the liquid formula diet for three days or a week, then a few days on the cottage cheese and fruit or all vegetable or all fruit or other diet of your personal selection.

As long as you keep your dieting at 900 calories or up to 1,200 calories or so a day, you're going to lose weight, and that's the objective.

Ten-Day Formula and Meat Diet

This bizarre diet which I worked out for some patients utilizes only the commercially prepared liquid 'formula foods' plus a meat dinner daily. It satisfies many people more than a one-dimensional diet. Follow this diet every day, no variations permitted, and you can lose weight quickly and quite pleasantly:

Breakfast: 1 glass of liquid formula

Lunch: 1 portion of formula soup or formula wafers

Dinner: 6 to 8 oz. of lean meat or poultry, moderate portion of a vegetable, moderate portion of fresh fruit or artificially sweetened fruit

At all meals you are permitted coffee and tea without milk, cream, or sugar, unlimited quantity of non-caloric beverages.

Temptation Diets

This 'newly beautiful' patient, Mrs A., had reduced in a few months from 160 to 130 pounds. She still had 10 pounds to

lose for her ideal weight of 120. Now she said to me pathetic-
ally, 'I'm thrilled about being slim again, Doctor, but I'm dying
for an ice-cream soda.'

'Fine, treat yourself to an ice-cream soda. In fact, you can
eat three ice-cream sodas a day for the next three days.'

She looked astonished, said, 'You're joking.'

'No. This is what I call a temptation diet. You yield to
temptation. You eat the ice-cream soda, or whatever particular
calorie-rich item you crave. You eat it three times a day for one
day or three days or even more. But that's just about all you
eat, except for a few low-calorie foods to make up your daily
total of 1,200 calories.'

'Won't I pile on weight in a hurry?'

'No. You'll still be consuming less than your daily caloric
requirements. Three ice-cream sodas will total 1,050 calories,
about 350 calories each, well below your daily requirement in
respect to reducing.'

Many doctors and nutritionists may deride this as a 'screw-
ball' dieting idea, but it works. I know it works because it has
helped slim many hundreds of people within my experience,
safely and effectively. By yielding to temptation this way, you
satisfy any craving for the calorie-rich food. In fact, after you
have three ice-cream sodas in a day, for example, let alone for
three days, you're so sick of ice-cream sodas that you may never
crave them again. And you realize that you really hadn't deprived yourself of anything wonderful, that there's much
greater reward in being thin than in being a glutton.

What's your temptation, if any? Is it ice-cream, pie, or just
plenty of fresh bread and butter? Do you crave a sweet baked
apple with cream, waffles or pancakes with butter and syrup?

Most dieters lose a craving after going on the Quick Weight
Loss Diet, or one of the drastic bizarre diets. But if the craving is really strong, let yourself go. *However, keep your intake
within your daily caloric allowance.* Don't add these temptation foods to the others; subtract them from the total.

Suppose your allowance for reducing is 1,200 calories a
day. (Stay Slim Eating usually runs much higher.) If your temp-

tation snacks, such as two ice-cream sodas, total 700 calories, you have 500 calories left for the balance of your day's eating. That means you can eat, for example, plenty of certain vegetables, a sizeable portion of salad with low-calorie dressing, a boiled egg, a thin slice of bread. And you can have all you want of non-caloric bouillon, sugarless coffee, tea, and non-caloric beverages.

Do you crave fruit pie? Have orange juice and coffee for breakfast, pie and coffee or tea for lunch, pie and coffee or tea for dinner, snacks of non-caloric bouillon and non-caloric carbonated beverages, as much as you want. You'll still total only about 1,200 calories per day (and you won't want to look at fruit pie for quite a while).

Is such eating dangerous to health? Absolutely not for a few days, a week, even a few weeks. Beyond that, a doctor should check you for any abnormalities in the blood or otherwise. As an extra measure, take a vitamin tablet daily. Don't forget that many clinics prescribe total abstinence from food for seven to fourteen days and longer. Many patients, under supervision, have had only orange juice and non-caloric liquids for sixty to eighty days, and taken off excess weight healthfully.

A woman complained to me, 'All my husband eats for breakfast is a scoop of ice-cream and black coffee, yet he stays thin and I get fat.' I said, 'Fine, the ice-cream has only as many calories as a dish of oatmeal. He's having about 150 calories for breakfast, that's very little.' I added up the foods she was eating and found that she was averaging an 800-calorie breakfast before she came to me. She reduced quickly with the Quick Loss method.

I certainly don't advocate ice-cream sodas and fruit pie, etc, as a permanent way of eating. Aside from caloric content, such rich foods may affect an acne condition or other skin problem, for example. But, as instructed here, yield to temptation on the temptation diet if it helps you to keep losing weight. Like most of my patients who have done it, you'll find that temptation soon gets behind thee.

Day-In Day-Out Diet

This simple 'lopsided' diet has served to reduce many patients in a hurry. You stick to the same menu day-in day-out for weeks. The very one-dimensional monotonous aspect of the eating is its most effective feature. As on many restricted diets, you lose your desire for much food after a few days. Total calorie intake is only 600 on the following:

> *Breakfast*: orange juice, black coffee without milk, cream, or sugar

> *Lunch and Dinner*: average portion of cottage cheese, a sliced tomato, 1 slice of bread, 1 teaspoon of jam
> Coffee or tea – as much as wanted throughout the day, without milk, cream, or sugar

Drink six to eight glasses of water daily and unlimited quantities of non-caloric carbonated beverages. One therapeutic vitamin tablet daily. If hungry, add one glass of buttermilk daily for 685 calories total.

Lop-Sided Egg Diet

For heavyweights who love eggs and should take off weight quickly, this lop-sided egg diet of only 450 calories has proved very effective:

> *Breakfast*: orange juice, coffee without milk, cream or sugar

> *Lunch and Dinner*: 2 eggs – soft-boiled, hard-boiled, poached or fried or scrambled in no-fat non-stick pan, 1 thin slice toast, coffee or tea, no milk, cream, or sugar

If this is too difficult for you, add about 100 calories a day with a half-grapefruit (75c), or a glass of skim milk (85), or melon (50–75), or a slice of toast with dietetic jam (100–125), or choose personal favourites of about 100 calories from the chart.

Drink lots of water and all the non-caloric carbonated beverages you want, and take a vitamin tablet or two daily.

Rice Diets

A rice diet is very simply made up predominantly of boiled white rice. This is not rice in its truly natural form since the shell or coating has been artificially removed. The best rice to use on these diets is the regular or extra long grain white rice, or par-cooked or par-boiled or converted white rice – the latter are all the same, but with different names. Pre-cooked rice costs up to three times more per pound than regular rice, has much of the flavour already cooked out of it and is less firm in texture.

Rice diets were originally introduced as an effective treatment for high blood pressure. They turned out to be fine reducing diets. Even though patients were permitted 2,000 or more calories a day, their weight kept going down, for two reasons. First, the rice diet has practically no salt. Second, because of the monotony of the food few people on the diet ate their full quota daily.

Here is the simplest rice diet. It will take off weight steadily and add years of life, particularly for those with a tendency to high blood pressure. It is not flavourful and it takes an individual with considerable willpower to stay with the diet week after week. In all the rice diets, no salt is to be used in cooking the rice or any other foods, nor may foods be salted at the table.

In all the following rice diets, drink as much as you want of water, coffee and tea without milk, cream, or sugar, and non-caloric carbonated beverages. NO fried foods are permitted. Take one or two vitamin tablets daily, particularly on diets No 1 and No 2.

Rice Diet, No 1

Every meal: 8 oz. fruit juice
3 oz. any fruit
$\frac{1}{3}$ lb. rice

This is a great deal of rice and it need not all be eaten nor is it usually (as noted, few people consume all the rice calories permitted on the diet). This same amount of food may be spread over six or more meals a day if desired. Take a therapeutic vitamin tablet daily.

Rice and Vegetables Diet, No 2

Every meal: 8 oz. fruit juice
3 oz. any fruit
Moderate portion of any vegetable
$\frac{1}{3}$ lb. rice

On this diet also you will lose weight steadily and consistently. Nothing is to be added to the diet. No salt or ketchup, just a little pepper or mustard or horseradish may be used, but always sparingly.

Rice, Meat, and Fish Diet, No 3

Every meal: add to Rice Diet No 2, a moderate portion of thoroughly rinsed lean meat or any fresh water fish, with no salt used in preparation or in eating. No seasonings with any hidden salt content, no salt or ketchup, a little mustard or pepper or horseradish.

Rice Variety Diet, No 4

Include some rice at every meal, and eat as much as you want of these foods: fruits, vegetables, well-washed lean meat, chicken, and turkey, fresh water fish, eggs, cottage cheese, unsalted matzos. *Nothing else is permitted — no salt used in cooking or eating.*

All these rice diets are excellent for taking off excess weight steadily. For those who are overweight and have hypertension (high blood pressure) these are literally lifesaving diets. The fact that the diets are especially good for an illness doesn't detract from their effectiveness when used specifically for reducing.

If you are not a hypertensive, after you are near or at your desired weight, you may add a little salt to the foods if you wish to make a rice diet your maintenance way of eating. Patients of mine have done this successfully and felt wonderfully healthy.

After losing all excess weight you may add in small amounts: sugar, treacle, honey; unsalted butter or margarine; jams and canned fruits which are not salted; bread when baked with fresh yeast only; white vinegar; spaghetti; peeled potatoes. Salt should be used sparingly if at all. If you start gaining weight, stop using the unsalted butter or margarine and substitute mineral oil for cooking and salad dressing (combined with white vinegar).

Always avoid: salted or smoked fish or meats — ham, bacon, sausages, herring, etc — pickles, ketchup, chili sauce, all such foods that have salt added in their preparation.

The average American consumes about 6 pounds of salt per year, and has been called over-salted. The rice diets, like many other drastic diets, eliminate much salt and water from the body, bringing down the weight.

All my patients on the rice diets lost weight, but none became skinny or scrawny. Self-denial is involved in staying on the rice diet particularly because the lack of salt makes food quite tasteless and removes the urge and desire to eat. None of my patients felt hunger pangs — it would seem that hunger pangs are associated with looking forward to a delicious meal! You may feel assured that on rice diets you will lose the strong desire to eat, and you'll lose plenty of pounds.

Choose-Your-Units Diet

This quick-reducing method pleases some dieters who like to make up their own daily menus from a choice of the whole variety of foods. Each 'unit' equals 100 calories of food, selected as you wish, from the calorie count listing in Chapter 7.

The fewer units you eat per day, the more swiftly you'll lose weight. For quick reducing I suggest a maximum of ten units (1,000 calories) per day. On six units (600 calories) you'll lose pounds and inches even more rapidly.

Example of a ten unit (1,000 calories) day:

Breakfast, two units:
 orange juice, one unit
 one egg, one unit
 coffee or tea, artificially sweetened

Lunch, three units:
 four-ounce hamburger, two-plus units
 tomatoes, under one unit
 coffee or tea, artificially sweetened

Dinner, five units:
 five shrimps, one-half unit
 baked or grilled fish, one-plus unit
 broccoli and carrots, one unit
 lettuce salad with lemon juice or vinegar, one unit
 melon, under one unit
 coffee or tea, artificially sweetened

Dieters have told me that playing the unit game adds a certain zest to reducing.

Six-Feedings-Unit Diet

Using the unit system, you will lose weight quite rapidly on twelve units (1,200 calories) a day by spreading the units over six feedings instead of the usual three meals, as follows:

Breakfast – two units
Mid-morning – one unit
Lunch – three units
Mid-afternoon – one unit
Dinner – four units
Mid-evening – one unit

As with all quick reducing diets, you may enjoy all you want of non-fat bouillon, coffee and tea (without sugar, cream, or milk) and non-caloric beverages.

900 Calorie Choice Diet

You will lose weight quite speedily if you stay within 900 calories a day on a varied diet. To help some of my patients do this I have saved them the trouble of consulting the long calorie listings by providing an abbreviated list (given here) under the name of the 900 calorie choice diet. Selected items and their caloric values are given for each meal. Following the listing is a sample day's eating such as you might choose for yourself.

Breakfast:
Fruits: apple 75 calories, cantaloupe 50–75, grapefruit 50–60, melon 50, 1 cup pitted cherries 70, peach 45–50, pear 95, 1 cup diced pineapple 75, 2 plums 60, 1 cup strawberries 55, tangerine 35, watermelon slice 1 in thick and 10 ins long 60

Fruit juices: $\frac{3}{4}$ cup grapefruit juice 65, $\frac{1}{2}$ cup grape juice 85, $\frac{1}{2}$ cup orange juice 55, $\frac{1}{2}$ cup pineapple juice 60, $\frac{1}{2}$ cup prune juice 85, 1 cup tomato juice 50

Bread: slice white 70, slice whole wheat 55, slice rye 55

Cereals: average serving of most unsweetened cereals about 100

Jams: 1 teaspoon about 20, 1 teaspoon dietetic 3–5

1 egg: poached, boiled, fried, or scrambled in fat-free pan 75

Skim milk: 1 cup 90

Lunch and Dinner:

Meats, lean: ¼ lb. 225 calories

Fish, lean: 6 oz. about 200

Shrimps, clams, oysters: medium each about 10

Chicken and turkey: boneless, no skin, 6–8 oz. 225

Cottage cheese: 1 tbsp. 15 calories, 4 oz. 135 calories

Vegetables: about 50 calories per sizeable portion of asparagus, broccoli, cabbage, cauliflower, celery, cucumber, eggplant, lettuce, peppers, radishes, sauerkraut, spinach, french beans, tomatoes

Salad: about 50 calories per sizeable portion dressed with lemon juice or vinegar

Starches: potato 95, ½ cup cooked rice 100, ½ cup noodles or spaghetti or macaroni 100, slice white bread 70 (no butter or sauces on any of these or on vegetables)

Now, from this condensed list, you know exactly what calorie value each item has and can make up your own daily menu such as this:

Breakfast: ½ cup orange juice 55, slice white toast 70, jam 20, coffee or tea without sugar, milk, or cream – total 145 calories

Lunch: meat 225, vegetable 50, rice 100, fruit 60 – total 435 calories

Dinner: cottage cheese 135, tossed salad 50, slice bread 70, melon 50, coffee or tea without sugar, milk, or cream – total 305 calories

Total for the day: 885 calories

While you don't lose weight nearly as rapidly as on the Quick Weight Loss Diet, unless you cut down considerably below 900 calories per day, this type of diet has one benefit. You start learning to figure your caloric intake almost automatically which is a help when you get down to your ideal weight and switch to Stay Slim Eating.

High-Fat High-Protein (HF-HP) Diets

Various diets high in fat and protein have been offered for some years. With this form of dieting the appetite for most people is well satisfied. There is a feeling of fullness and of eating a lot, with no accompanying 'hunger pangs' (which are more often felt in the dieter's mind than in his stomach). Thus the job of dieting is made easier. However, for speedy reducing which I advocate strongly, my Quick Weight Loss Diet and others in this book are far more effective.

Because they help some people to reduce I include my version of HF-HP diets which I have found to be understood most readily and to work best in this category. Success depends on adhering strictly to the regulations; deviations may well result in no weight loss. It is a fact that through the years such famed explorers as Wilhalmut Steffenson and Admiral Perry, Eskimo, and the Indians of the Northwest Plains lived primarily by this way of eating. They were able to perform Herculean tasks and were rarely heavy.

Of course their bodies were exercised considerably in their daily energetic routines. This is an integral factor in all dieting since exercise helps to firm soft, unused muscles, helps prevent sagging and burns up some calories.

In the HF-HP diets you avoid eating carbohydrates (sugars, starches, celluloses) and you utilize the stored fat in your body. The usual basic restriction at its strongest may be summed up in one sentence: *Don't touch carbohydrates, sugar, salt, fruits, cereals, bread, rice, potatoes, alcohol.* Much food is permitted but there are also many prohibitions. Use of salt is highly restricted or not permitted at all.

HF-HP Diet No 1

Eat plenty of the following, three times a day if you wish:

Any non-smoked and non-salty fatty meats, chicken, duck, fish, shellfish. Leave fat on meats and poultry butter, margarine, and oil as desired. Fry fish and shellfish in heavy oils.

Eggs fried in butter, margarine, or oil.

Cream cheese, cottage cheese, sour cream, heavy cream.

Don't use any salt.

Drink plenty of water, non-caloric beverages, artificially sweetened coffee and tea with cream if desired but no milk.

No snacks are permitted. Eat plenty at each meal, as much as you can without overeating.

No fruits, vegetables other than lettuce and tomatoes, no breads, spaghetti, macaroni products, rice, potatoes, cakes.

No pickled or canned meats or fish, no frankfurters, bologna sausage, salami, smoked or salted ham or bacon (unsalted ham and bacon are permitted).

No alcoholic beverages, no beer, not even dry wines.

A sample day's eating from the foods allowed:

Breakfast:
  2 scrambled eggs fried in butter or margarine
  3 strips of unsalted bacon
  coffee with cream (not milk)

Lunch:
  6 oz. fish or shellfish fried in oil
  sliced tomatoes with oil or mayonnaise
  coffee with cream

Dinner:
  8 oz. of fatty meat
  salad with plenty of oil or mayonnaise
  coffee with cream

Snack 10 PM:
  cheese, or chicken, or hard-boiled egg

Make up your own daily menus from the foods permitted, along the lines of this sample day's eating, on this and the following HF-HP diets.

HF-HP Diet No 2

This is the same as Diet No 1 but you add a cup of any one of the following vegetables at lunch and dinner if you wish:

Asparagus	Celery	Lettuce
Broccoli	Cucumber	Radishes
Brussels sprouts	Eggplant	Tomatoes
Cabbage	French beans	
Cauliflower	Green pepper	

HF-HP Diet No 3

This is similar to Diet No 2 but you may now add some of the following vegetables (1 cup at lunch and at dinner of any one) which contain carbohydrates in a higher percentage:

Beans, baked	Peas
Carrots	Potatoes
Corn	Potato chips
Onions	

HF-HP Diet No 4

This is the same as Diet No 3 but here you may add:
   Half-slice of protein bread three times daily.
   Half grapefruit or a small orange each day.

General Instructions and Summary for HF-HP Diets

Don't try to gorge yourself at the beginning of your dieting as the abundance of fatty foods may tend to make you nauseous.

You may eat as much fatty food as you wish within the restrictions of the varieties permitted. Eat to the limit of your capacity at each meal but if you overeat you are likely to feel ill.

No smoked or salted foods, but water-packed salmon and tuna are permitted.

Don't add salt to any foods. You may season with pepper, mustard, and dry spices.

Eat as much protein as you wish but only from the foods permitted. The proteins will come from animal fats and pure oils primarily; proteins in carbohydrates are generally forbidden.

Drink six to eight glasses of water daily to wash out acids and waste in your blood. Drink plenty of water before, during, and after meals. Without enough water there may be a tendency to become listless and tired on this type of dieting.

In general you may use any type of oil or fat or butter on these diets. However it is desirable, for men especially, to use unsaturated fats or margarine containing a great deal of corn oil, safflower oil, or soybean oil.

If you have been advised that you have a high cholesterol, trim your meats of fat. Substitute plenty of margarine for frying, and instead of gravies.

There is very little residue from this type of dieting so don't expect the usual frequency of bowel movements. If you feel constipated take a mild laxative such as milk of magnesia.

The body uses your own or the eaten fats as fuel on HF-HP diets. The sugar normally produced by the liver is generally sufficient to supply quick energy as needed. By drinking plenty of water as advised, you should feel sufficient energy.

Test your urine with *acetest* (may be purchased at chemists). A drop of urine on this white powder will turn purple if you are following the regulations correctly.

Exercise in the form that suits you best for thirty minutes continuously each day (before breakfast is a good time for a brisk walk, for example).

Don't sleep more than eight hours of each day.

Take a vitamin pill daily.

Grapefruit-Plus Diet

This diet is satisfactory to many. It combines the restricted foods for extra efficiency in burning up accumulated fat. It provides plenty of food and is a relatively easy diet to maintain if you like grapefruit very much. It becomes most effective in weight drop after five days, then you will lose steadily and quite rapidly although not as fast as on the Quick Weight Loss Diet. Follow this menu daily, varying the meats and salads:

Breakfast:
    half grapefruit or 4 oz. unsweetened grapefruit juice
    2 eggs any style
    2 slices bacon
    coffee or tea

Lunch:
    half grapefruit
    meat, poultry, fish, or shellfish – moderate portion
    salad, as much as you want with low-calorie dressing or lemon juice or vinegar
    coffee or tea

Dinner:
    half grapefruit
    meat, poultry, fish or shellfish – moderate portion
    salad, plenty with low-calorie dressing, lemon juice, or vinegar
    coffee or tea

With this diet you may not eat anything between meals; at bedtime you may have four ounces of tomato juice or skim milk. No desserts or alcoholic beverages are permitted. Enjoy as much as you want of artificially sweetened carbonated beverages, also artificially sweetened coffee or tea without cream or milk.

Two Meals a Day Diet

This is a highly restricted, simple diet which has proved very popular and successful with a number of my overweight patients. It is easy to remember and takes off a pound a day for the average overweight individual, and 2 to 3 pounds a day, day after day, for those who are very heavy.

The diet consists of two meals daily. Each meal is a repetition of the other, aside from your choices within each variety permitted. Eat only what is prescribed, nothing else. Since you have a wide choice within the restrictions, each meal can be enjoyed – especially as you see your weight on the scale dropping away every single day. Each meal consists of:

> 3½ oz. lean meat, poultry, fish, or shellfish
> 1 cup of a low-calorie vegetable
> ½ slice thin toast
> 1 apple or orange, small
> coffee or tea with a tablespoonful of milk if desired, and non-caloric sweetener
> no oil or butter

You may use non-caloric salad dressing, lemon juice or vinegar.

Use salt and pepper in moderation.

The juice of one lemon is permitted for the entire day, may be taken as lemonade made with artificial sweetener.

No substitutes allowed on this listing.

The calorie intake on this diet is 250 per meal, 500 for the day.

Have your meals at whatever hour you wish, only two meals per day. After a few days dieters find that their weight keeps dropping steadily and rapidly.

Stay on this diet for twenty-one days, or stop it earlier if you've reached your ideal weight. Then switch to Stay Slim Eating for two weeks. If you still have not attained your ideal weight, return to this two meals a day diet for two more weeks. Continue the alternating schedule until you're at your ideal weight. Then proceed with Stay Slim Eating. Repeat this cycle in the future as often as may be necessary.

Do not expect the normal frequency of bowel movements while on this diet because of the lack of bulk ingested. One movement every three or four days will be normal, especially if you drink plenty of water which is highly desirable (six to eight glasses a day). You may use a mild laxative, milk of magnesia or a glycerine suppository as an aid.

The results of this diet are extremely rewarding. With my patients, notable results have included upsurge of energy, loss of breathlessness, lessening of excessive perspiration, relief from high blood pressure, lowering of cholesterol, blood sugars, and uric acid. In a number of cases irregular menses have become normal, sex desire increased and fertility improved.

Many undesirable signs and symptoms associated with over-weight have improved considerably. I've had many patients drop over 30 pounds in three weeks on this diet with no ill effects and with a most gratifying improvement in health and an uplifting feeling of well-being.

Anti-Acne Diet

While this dieting method is designed especially to help relieve and control an acne condition (which it accomplishes with considerable success), it is also an effective quick reducing guide. While eating properly is not the cure-all for acne, it is an effective and usually essential part of helping to prevent or clear up the condition.

With the anti-acne diet you eat as much as you need to satisy you, *not* stuffing yourself. Enjoy plenty of proteins – lean meat, poultry (with skin removed), lean fish, and shellfish. It is of prime importance *not* to eat certain foods, specifically the following:

No fats, not even those in butter, margarine, cream, whole milk, and whole milk cheeses. You can however enjoy creamed cottage cheese and other skim milk cheeses.

No chocolate or other sweets containing sugar; no cocoa.
No fried or highly seasoned foods.
No mayonnaise or rich dressings.
No pies, rich cakes, pastries, icings.
No pizza or anything with a rich crust.
No sugar-content minerals or other sugary beverages.
No alcoholic beverages.

There's a great deal you can eat and drink with enjoyment. Drink plenty of sugar-free beverages, fruit juices without sugar added, coffee, tea. In addition to all the delicious lean meats, poultry, fish, and shellfish (use lemon or vinegar instead of mayonnaise), enjoy fresh fruits and vegetables (no butter or creamy sauces – use lemon juice), artificially sweetened jelly, and other sugar-free desserts.

If you have an acne condition your skin is likely to show gratifying improvement on this diet, as your figure slims down. The Quick Weight Loss Diet is also helpful in helping to combat acne.

Anti-Gout Diet

This diet is a boon to those who are afflicted with gout or have a tendency to suffer from it. Of course with this ailment you will be guided by your doctor's overall instructions. This diet has been extremely helpful to my patients suffering with gout, and is offered subject to your physician's approval in respect to your own case.

People who suffer from gout usually have an excessive amount of uric acid in the blood. They are generally heavy and are apt to eat excessively, particularly of the wrong foods. On this diet, meats and other foods high in purines (uric-acid compounds) are permitted in very limited quantities; in some cases meats are not allowed at all. Choose your daily intake from this listing:

Skim milk – 2 glasses daily.
Lean meat, poultry, fish, shellfish – no more than 4 oz. daily.

Avoid entirely sweetbreads, sardines, anchovies, salmon, tuna, smoked fish, liver, brain, kidney, meat extracts, all glandular foods.

1 egg daily.

Vegetables – green, yellow, and leafy – substantial portions without butter or sauces. Eat sparingly or avoid lentils, beans, peas, spinach.

Potato, rice, macaroni – small to moderate portions.

Fruits – half grapefruit, orange, or other citrus fruit, twice daily.

Toast or bread – 3 slices daily. Eat sparingly or avoid whole wheat bread, porridge, whole wheat, or heavy cereals.

1 pat of margarine or butter daily, used as desired.

Desserts – no chocolate or nuts. Artificially sweetened jelly and citrus fruits are permitted.

No salad dressings; vinegar or lemon juice are permitted.

No soups made with meat extracts.

No coffee, cocoa, or alcoholic beverages.

Sample day's diet:

Breakfast: fruit or juice
1 small portion light cereal with skim milk
1 slice enriched white bread (toast)
plenty of very weak tea, 1 tea bag to $\frac{1}{2}$ quart of boiled water; drink at least 3 cups

Lunch: vegetable broth (no meat extracts) or 1 egg
potato, rice, or macaroni
salad with vinegar dressing or lemon juice
1 slice white bread and pat of butter or margarine
1 glass skim milk

Dinner: vegetable broth (no meat extracts)
3 to 4 oz. meat, chicken, fish, or shellfish
potato, rice, or macaroni
salad with vinegar dressing or lemon juice
1 slice white bread
3 to 4 cups of weak tea

Drink plenty of water all day, at least two quarts. One glass of skim milk mid-evening. Drink as much seltzer and artificially sweetened beverages as desired.

With this Anti-Gout Diet you can still enjoy plenty of delicious, satisfying foods daily from the listing. For those who suffer from gout this method of eating will help provide relief as the pounds and inches slip away and general health and well-being improve. It must be repeated that eating is a vital part but only a part of overall anti-gout treatment.

Anti-Bloat Diet (Anti-Water Retention)

This diet is especially helpful for the overweight who feels not only heavy but also bloated much of the time. This reducing method promotes an acid condition which stimulates greater amounts of water (urine) excretion. It brings remarkable relief to bloated persons with swollen joints, fingers, feet, and legs, particularly in the summer months. It also helps those with phlebitis or varicose veins.

Each day you can enjoy foods in the quantities specified in the following listing, making up your own meals accordingly:

8 oz. milk

two 4-oz. portions of lean meat, poultry, fish, or shellfish

1 egg

2 half-cups of vegetables; corn is permitted, asparagus especially recommended

1 medium-size potato

4 oz. unsweetened orange juice

3 slices of toast

1 portion of cereal

cheese, moderate portion

fruits – cranberries, plums, or prunes should be eaten daily; for quicker weight loss drop the toast and cereal daily and substitute artificially sweetened cranberries or prunes or fresh plums.

Anti-Allergy Diet

For some overweight patients who are subject to a variety of allergies, I have created these anti-allergy diets. Individuals lose weight rapidly on this routine and frequently obtain some relief from their allergies at the same time (of course the diets did not replace specific investigations of the basic causes of the allergies). It helped to keep these overweights reducing by providing A- and B-day diets. Follow A for three days, then B for three days. If allergies subside, alternate A and B by days, otherwise continue three-day cycles.

A-Day Diet:

Breakfast:
   half grapefruit
   coffee or tea, artificially sweetened, no cream or milk

Lunch:
   salad of 1 tomato, $\frac{1}{4}$ head lettuce, 5 asparagus tips, lightly
      dressed with vinegar or lemon or garlic
   cold lean lamb or beef – 2 small slices
   1 pear, medium size
   coffee or tea, artificially sweetened, no cream or milk

Dinner:
   cup of clear broth
   lean lamb, beef, or chicken – 1 slice 4 ins. × 4 ins. × $\frac{3}{4}$ in.
      thick
   spinach – $\frac{1}{2}$ cup
   carrots – $\frac{2}{3}$ cup
   large artichoke with vinegar and mineral oil dressing
   1 peach, medium size
   coffee or tea, artificially sweetened, no cream or milk

B-Day Diet:

Breakfast:
  half grapefruit
  1 piece rye crispbread
  bacon, 3 slices, very lean and crisp
  coffee or tea, artificially sweetened, no cream or milk

Lunch:
  same as on A-Day diet, but add 1 piece rye crispbread

Dinner:
  same as on A-Day diet, but add 1 medium size potato
  and add a cup of sugar-free or artificially sweetened pineapple instead of the peach

On both diets add a pill of vitamins and minerals daily. You may have all the artificially sweetened carbonated beverages and soda you want. Drink plenty of water daily.

Stay on this diet for four to six weeks, or stop earlier if you reach your ideal weight. Then switch to Stay Slim Eating for two weeks, or from then on if you're at ideal weight. If necessary to reduce further, return to the A-Day/B-Day schedule or go on the Quick Weight Loss Diet.

Selective diets for specific ailments and conditions, and for unhealthy, diseased, or malfunctioning organs, are generally advocated now. They have been helpful in many cases. Such diets for various diseases or afflictions are given because scientifically, experimentally, or through results in practical experience, they have helped relieve chemical deficiencies or excesses.

Foods that may cause or aggravate illness in some specific instances are omitted or reduced in these diets. In respect to the diseases mentioned here, applications and usage of different foods have some relieving and curative effects.

Since each human is a complicated piece of machinery,

many factors may affect his well-being. Some of the suggestions for the use of these general diets are given simply to supplement your doctor's orders which should be followed carefully. I have no intention at any point in this book to treat anyone's specific ailments but to be of aid by explaining to you how your doctor's diet and other diets may be helpful.

Anti-Rheumatoid Arthritis Dieting

Rheumatoid arthritis is a diffuse (not concentrated) chronic disease which affects the muscular and bony tissues especially around the joints. During the course of the disease there are usually periods of severe pains. These are due primarily to the retention of inflammatory fluids around the tissues and joints. Your doctor can help you counteract the pains of the condition and should be consulted periodically.

With my patients I recommend the intake of foods high in vitamin B complex; and those with an acid ash content low in sodium (salt) and low in carbohydrates. Eating this restricted type of food usually helps in decreasing the joint and muscle swellings. Some of the foods recommended:

Acid Ash Foods: cranberries, eggs, fish, meats, oysters, plums, prunes, refined rice
Low-Carbohydrate Foods: artichokes, asparagus, cabbage, cauliflower, cucumbers, lettuce, radishes, spinach, brussels sprouts, french beans, tomatoes
Low-Salt Foods: beans, cabbage, cranberries, grapefruit, grapes, lemons, plums, watermelon
High B-Complex Foods: whole wheat, wheat germ

Typical Day's Menu

Breakfast:
   1 slice whole wheat bread, toasted
   1 poached egg
   tea or coffee

T—E

Lunch:

4 oz. meat, chicken, or fish, or 1 egg
½ cup refined rice
½ sliced tomato
tea or coffee

Dinner:

vegetable soup
2 oz. meat with ½ cup refined rice
½ grapefruit
tea or coffee (plentiful)

In order to make up the high B-complex needs use an all B vitamin capsule and sprinkle wheat germ over some of the food.

The daily menus can be varied as you wish from all the foods allowed. Eat only limited quantities that will bring you down to ideal weight and keep you there. Overweight is extremely undesirable with this affliction, as with most other illnesses.

Raw Food Diet

This diet was extremely popular in Europe at the end of the last century for the relief of rheumatic aches and pains. Excellent results were reported, especially in helping the overweight.

No cooked fruits or vegetables were allowed, only raw. The diet provided sufficient eggs and cheese so that there was minimal water retention.

In this dietary format, fruits and vegetables should be raw but it is permissible to liquefy them in a blender or by other means. While I cannot agree with the curative values claimed at the time of its peak popularity, there's little question that if the diet is followed, within the set limitations, there will be a constant weight loss down to ideal weight. There should also be relief from the severe pains and annoyances usually accompanying this affliction.

You can make up your own daily menus, from the specifications already given, for this type of eating.

Breakfast:

  raw fruit of your choice, unsqueezed

  1 poached egg

  1 thin slice whole wheat bread

  tea, coffee, water (plentiful)

Lunch:

  1 or 2 cups of raw vegetables, cut up if desired

  1 piece of raw fruit of your choice

  1 to 3 bread sticks, salt-free

  1 pat of unsalted butter

  tea or coffee

Dinner:

  2 eggs poached, boiled, or hard-boiled (no fat)

  moderate portion of cheese

  piece of raw fruit

  1 slice of toast with jam

Snacks:

  4 PM and 10 PM – raw fruit, or raw vegetables, or a few unsalted bread sticks

In this anti-arthritis diet, as in any diet, always try to avoid foods to which you may be allergic or which generally produce uncomfortable gastric symptoms.

If this diet helps you feel better, as it does many of those afflicted, you may stay on the diet for six months, but always make sure that you take sufficient eggs and cheese along with the other recommended foods (with the approval of your physician, of course).

Anti-Osteoarthritis Dieting

The Osteo type is usually a growing-old arthritis. It rarely cripples like rheumatoid arthritis but its accompanying aches and pains can make life miserable for the afflicted. Common symptoms are neck and shoulders pains, headache, and neuralgia, severe backaches, pains in the pelvis, groaning pain in the knees and ankles. There is a tendency for the person to be 'always tired'.

If you have these symptoms and they are diagnosed as osteoarthritis, you will certainly be under the guidance of a physician and will follow his advice.

Overweight aggravates and hastens this condition rapidly, increasing the pains and suffering. Heavy patients in my care are instructed to take off excess weight quickly; reducing as fast as possible takes the extra load off the aching body, legs, and joints.

Here is the regimen I recommend to my patients (instructions for the diets are given in preceding pages):

1. Start with one to two days on a fasting diet.
2. Then go on the Quick Weight Loss Diet.
3. Stay on the QWL Diet until down to ideal weight.
4. Keep weight down with Stay Slim Eating.

Invariably when the weight starts to come off, the aches and pains decrease. Great relief is usually obtained when the individual is at ideal weight and never allows the excess pounds to pile on again.

Anti-Kidney Stone Dieting

Many excessively overweight individuals suffer from kidney stones. Some of the stones are dissolved relatively easily through diets and medicines. Visits to a doctor, and following his instructions to the letter, are mandatory.

The counteracting aim in treating all stone depositing is to reduce the tendency towards calcium deposits. If the condition calls for the urine to be acid (you will be guided by your physician, of course), avoidance of all milk and cheese is basic and essential. A combination of high acid ash foods and low calcium foods is necessary.

Those patients who have uric acid stones or gravel are usually given an eating regimen which is just the reverse. This involves intake of a minimal amount of protein (meats, etc) and an alkaline ash diet.

To make certain that the diet results in urine that is either high alkaline or high acid (whichever is called for in the particular case) a frequent litmus paper test of the urine is essential. In these tests, a red litmus paper turns blue as an indication of alkaline urine. With blue litmus paper, if it turns red the indication is an acid urine. This simple testing is usually done three times daily to get the most dependable results and guidance.

It is significant that with patients afflicted with stones, common findings are overweight and the drinking of too little water.

For patients who have uric acid stones or gravel, a diet like the following is often recommended. This is an alkaline diet and a low-meat diet. Here is a typical day's menu under 900 calories, making it an effective reducing routine:

Breakfast:

    half-grapefruit or 1 sliced orange
    1 egg
    1 slice white toast and dietetic jam
    tea or coffee and water (plentiful)

Lunch:

    5 tbsps cottage cheese
    1 sliced tomato
    1 slice white toast and dietetic jam
    tea or coffee

Dinner:

> 4 oz. lean meat
> 1 cup of vegetable except spinach or brussels sprouts
> 1 slice pineapple
> 1 slice white toast

Snack:

> grapefruit or orange, mid-afternoon or mid-evening.
> Drink plenty of water with meals, all day and evening.
> Be careful to avoid eating cranberries: plums, prunes, spinach, brussels sprouts.
> Only vegetable soups (no beef stock) are permitted.
> No alcohol is allowed.
> These high-purine foods are omitted, the same as in treating gout: anchovies, biscuits, herring, gravies, meat extracts, organ foods (brain, kidneys, liver, sweetbreads), sardines.
> Not more than four to a maximum of six ounces of meat are allowed daily.
> With this type of diet, as with most diets, drinking water 'plentifully' means six to twelve glasses of water daily. This is in addition to any other beverages imbibed.
> *For patients who should have acid urine*, the recommendation is for a diet low in calcium, low in phosphorus, high in acid ash. It is suggested that the dieter take a great deal of tea and coffee plus all the water he can drink; taking over ten glasses of water a day is highly desirable.
> This diet should be high in chicken, eggs, fish, meats. General eating should include canned and cooked fresh fruits, cranberries, plums, prunes.
> Avoid grapefruit and oranges.
> Omit carbonated beverages because they tend to be alkaline.
> Omit all rye, wheat, or whole grain breads. Plain processed enriched white bread is permitted.
> Omit any ordinary or special whole grain cereals.
> Refined rice is permitted.
> Very few desserts except fruits and jelly are allowed.

A small amount of salad dressing made from eggs is permitted.

No milk.

No leafy green or yellow vegetables.

No nuts or yeast.

Here is a typical daily menu which may be selected according to personal liking from the permitted foods:

Breakfast:
   half-grapefruit
   1 egg
   1 slice white toast or portion of refined rice
   dietetic jam
   tea or coffee

Lunch:
   4 oz. lean meat
   moderate portion noodles or spaghetti or macaroni
   1 sliced tomato
   1 slice white bread with dietetic jam
   tea or coffee

Dinner:
   2 oz. lean meat
   moderate portion of carrots
   moderate portion of refined rice
   moderate portion cranberries
   cooked fresh peaches
   tea or coffee

Snack:
   plums, mid-afternoon or mid-evening

This type of dieting is low in calories and qualifies as an effective quick-reducing regimen.

Anti-Hyperhidrosis (Sweating) Dieting

Hyperhidrosis (excessive sweating) can make a person miserable and should be quickly counteracted. It occurs primarily in the excessively overweight individual (there may be other considerations too, of course). The water (perspiration) almost pours from the very heavy person under many conditions.

Every crevice in the body thus becomes moist and macerated (tender, soft, pulpy). The most trouble-prone areas are usually under the breasts, in the groin and the insides of the thighs. This condition is a source of almost constant suffering for the afflicted woman or man.

The first and basic course of action is to get that excessive weight off rapidly. My patients are given this regimen:

1. Start with a one or two day fasting diet.
2. Then go on the Quick Weight Loss Diet.
3. Stay on the QWL Diet until down to ideal weight, varying along the way with other quick-reducing diets if necessary.
4. When at ideal weight, stay that way permanently by Stay Slim Eating. Go back on the Quick Weight Loss Diet immediately if weight increases more than 3 pounds over ideal weight.

The individual afflicted with hyperhidrosis should switch to cool, light clothing. Powder all tender, affected areas frequently, powdering the entire body whenever convenient.

Once down to ideal weight, and well launched on Stay Slim Eating, the condition usually vanishes.

Anti-Skin-Troubles Dieting

For any kind of skin trouble it is vital to consult a dermatologist before the condition becomes worse. With excessive weight the skin generally is more vulnerable to excess perspiration with its attendant problems of oiliness, inflammation, chafing, prickly heat, and fungus infection. Again it's highly desirable to get those excess pounds off quickly.

It has been found that in many cases excessive fatty acids in the skin and in the blood predisposes the individual to certain types of growths on the skin and eyelids. These solid reddish and brown papules (swellings, pimples) may also be found over the body (the lesions around the eyes are usually known as chloasmas). A diet low in fat, and especially low in cholesterol, followed over a long period of time may help relieve and clear up this condition to some extent, if not entirely.

Follow the diets previously given in this book, particularly the diets which allow only a minimal amount of cholesterol and fatty acids. The diets attack the primary cause for a good many people by bringing down weight and alleviating skin problems perhaps due to the excess body fat.

Anti-Psoriasis Dieting

Those persons who are afflicted know too well what psoriasis is – a chronic condition involving erythematous (red) patches generally, which are usually covered by whitish scales. It may take different forms in individual cases. A physician and dermatologist should be consulted and instructions followed to the letter.

While there is no known cure at this time for psoriasis, many doctors believe that the condition generally involves excessive oiliness of the skin and in the blood. A diet which is very low in fat, fatty acids, triglycerides, and cholesterol is often helpful to the afflicted.

Psoriasis is often associated with other fat disturbances such as oily seborrhea (dandruff), gallstones, osteoarthritis, and overweight.

Reduction of weight is essential as one step in combating psoriasis. The low-fat diets given in preceding pages are usually recommended and are often very helpful in obtaining some relief.

Anti-Dyspnea (Shortness of Breath) Dieting

Dyspnea is the medical term for difficult or laboured breathing, commonly described as shortness of breath. At the first sign of any such difficulty, which may derive from a variety of causes, the afflicted individual should see a physician and be guided by him.

With most patients in my care, I find that the Quick Weight Loss Diet is especially helpful when excessive weight is involved, and it often is. It not only reduces the overweight person quickly, but it also helps to draw excessive water out of the tissues and blood.

After getting down to ideal weight on the QWL Diet, the individual has the benefits of a smaller liver, a smaller stomach, reduction of fat around the heart, reduction of fat over the entire chest walls, and reduction of the blood circulation through the lungs. All these changes tend to lessen the breathing problem. Shortness of breath due to excessive fat (or other factors) is a cruel affliction.

This is the course I recommend to my overweight dyspneic patients:

1. Start the Quick Weight Loss Diet *at once*.
2. Stay with this diet faithfully until down to ideal weight.
3. Then go on Stay Slim Eating for the rest of your life.

Patients who follow this routine find a wonderful new freedom in breathing and healthier living. They rarely suffer again from such difficult, laboured breathing.

Anti-Acid Diet and Anti-Ulcer Diet

When the stomach secretes too much hydrochloric acid, it produces an hyperacidity which in turn results in heartburn, sour taste, a bloated feeling, and even pain in the pit of the stomach. In this condition, most of the spices, condiments, and herbs which tend to inflame the gastric mucosa should be avoided.

All stimulants should be reduced, including alcoholic drinks, beer, tea, coffee, cola drinks, even aspirin products. Foods with concentrated extracts also generally irritate the stomach.

An ulcer of the stomach may be involved. This is an aggravated inflamed mucous membrane. Other factors such as heredity may enter into a tendency towards a stomach ulcer. Primarily it involves a very circumscribed small area which ulcerates and becomes very painful. Immediate and continuing physician's care is essential with such a condition.

It is helpful to eat only bland foods, without spices. The stomach juices must be neutralized. This requires eating often. There are two generally accepted eating routines which I recommend to many of my patients with this condition:

1. Drink skim milk every two hours, with a little refined rice or a few water biscuits.
2. Go on the Quick Weight Loss Diet at once, spreading eating out over the day, with only a few mouthfuls eaten every one and a half to two hours.

My patients who follow either form of dieting until relieved do not gain any weight. In most cases they lose weight rapidly when overweight, which is a great help in itself for the general improved physical condition.

Once an individual develops hyperacidity, whether from the type of food eaten or the large quantities ingested, he is almost always subject to recurrences unless correction is undertaken quickly and maintained. Persons in my care invariably remain free from recurrences by following the QWL Diet until down to ideal weight, and continuing to stay at that plateau. They are then likely to maintain good health, free from pain and discomfort; the same is true with any sensible, bland, soft, low-calorie way of eating.

Anti-Proctitis Dieting

Proctitis is the medical term for an inflammation of the lower bowel and the anal region. There are many local and systemic causes which your physician will investigate and treat accordingly.

Usually eating too much of highly spiced foods and drinking too much alcohol are involved. In the lists of diets given here, the Quick Weight Loss Diet or any of the bland diets provided will usually relieve this problem in a few days unless some special condition is involved.

It is vital here again that the afflicted individual get down to ideal weight and then maintain that status. For this purpose the Quick Weight Loss Diet is recommended primarily, followed later by Stay Slim Eating.

Diversion Diet

Occasionally a patient who has been reducing rapidly and steadily on the Quick Weight Loss Diet will come to me and say, 'Doctor, I'm dying for something creamy. I must have something rich, not sweet but rich and creamy. What can I do?'

I have just the diet for this situation. I tell the patient to go on my diversion diet which I developed especially to satisfy that urge to eat 'something creamy, not sweet but rich and creamy'. I suggest that the overweight person eat nothing but the following for the next two or three days:

5 tbsps cottage cheese mixed with $1\frac{1}{2}$ tbsps sour cream. Eat this dishful six times a day at whatever hours you wish.

Also, you may have a cupful of chicken or beef bouillon (no fat), hot or cold, with each meal.

In addition, you may have as much as you wish of tea and coffee, hot or iced, artificially sweetened if desired, but without milk or cream.

Enjoy unlimited quantities of artificially sweetened carbonated beverages each day.

In case after case the patient finds that the diversion diet hits the spot, is 'just what the doctor ordered'. It satisfies the craving. The individual keeps losing weight rapidly each day. The reason is that the cottage cheese and sour cream mixture totals 110 calories per meal, only about 660 calories per day.

After two or three days the overweight has satisfied that particular craving thoroughly (or may stay on the diet a few days longer if desired). It usually meets the need for, as one woman said, 'It is something that tastes like 2,500 calories but only adds up to about 700.'

He or she is then ready to go back on the Quick Weight Loss Diet with its plentiful variety of good foods and the swift and gratifying drop each day of the numbers on the scale.

Bread-Cheese-Wine Diet

This diet has proved successful with certain patients who like wine and cheese very much and don't wish to give up alcoholic beverages completely while taking off excess weight rapidly. Extended over a period of weeks the diet has worked out well in such cases. The only variety is in the wide choice of cheeses and wines. Only light wines should be used, not the heavier types such as sherry and port.

You might wish to try this unique method of dieting for a week, then switch to the Quick Weight Loss Diet which permits you a greater variety of foods yet takes off pounds and inches most efficiently. Your goal is to take off excess, burdensome weight, and if you miss wine very much, this may be just the diet that works best for you for a period of a few weeks at a time.

Breakfast:
   1 slice toast
   4 oz. cheese (your choice)
   coffee or tea, artificially sweetened, no cream or milk

Lunch:
2 slices rye bread (or your choice)
   4 oz. cheese, your choice such as 1 oz. Roquefort, 1 oz. Camembert, 1 oz. Swiss, 1 oz. Cheddar

4 oz. wine (your choice of light wines)
coffee or tea, artificially sweetened, no cream or milk

Dinner:
same as lunch

The preceding is recommended for rapid reducing. If desired, add another slice of bread and 3 oz. wine as a mid-afternoon or mid-evening snack, but weight loss won't be as quick.

Three-Week 555 Diet

This stringent diet has worked well for some individuals who have great determination to lose weight fast and can stick to a rigid routine for three weeks. This is called the three-week 555 diet because the daily calorie allotment totals about 555:

Breakfast:
4 oz. orange juice or tomato juice
coffee or tea, artificially sweetened, no cream or milk

Mid-morning:
1 glass skim milk

Lunch:
lettuce, large portion with salt, lemon juice, or vinegar
1 egg (any way, no butter)
coffee or tea, artificially sweetened, no cream or milk

Mid-afternoon:
1 medium-size apple

Dinner:
large salad with half-and-half vinegar and low calorie
    dressing
½ slice thin toast
cup of bouillon
3 oz. lean meat or chicken without skin
coffee or tea, artificially sweetened, no cream or milk

Have a vitamin-mineral pill daily. As much artificially sweetened carbonated beverages as you want. Drink plenty of water daily.

After three weeks, or if ideal weight is reached earlier, switch to Stay Slim Eating for a few weeks. Then, for further reducing, return to the three-week 555 diet or go on the Quick Weight Loss Diet.

The Quick-Loss diet is my prime recommendation as the fastest, surest, most desirable quick-reducing method for the great majority of overweights.

Take Your Choice . . . or Choices

Certainly from the Quick Weight Loss Diet and over sixty other quick reducing diets here you'll find one or more just right for you. If you reach the point where you can't stay on your selected diet, switch to another. For example, after a week or more on one diet, if you have a yearning for vegetables and fruit, go on the vegetable and fruit mix diet. If you want meats, go on the meat only diet. They'll all keep bringing down your weight.

Once you start on a stringent bizarre diet, you may be amazed as I have been so often, at the depth of willpower we humans possess. I've had a patient choose to eat nothing but ¼ lb of meat with lots of sauerkraut daily for weeks. Others chose as their personal rigorous diets raw chop-meat only, ½ lb of cottage cheese mixed with ketchup daily for fourteen weeks, 1 hard-boiled egg every third day for two weeks.

These determined dieters lost a great deal of weight, felt well and vigorous. My check-ups showed them to be in good health. I'm not suggesting such harsh diets to you. I cite them as examples of what individuals will and can do to get down to desired weight once they're sufficiently motivated. You can certainly accomplish needed weight loss for yourself through your choice of the many quick reducing diets here that are much easier to follow than a single-food routine.

In following many of the bizarre diets, here are a few essential guides:

1. Quick reducing diets are not recommended here for growing children (although I have treated hundreds of overweight youngsters effectively by rigorous dieting), or for adults with specific ailments such as diabetes or kidney, liver, or heart disease. Such persons should be under the care of a physician, of course, who will attend to their reducing as well as other matters. These diets are not intended primarily to cure specific illnesses, particularly critical and serious ailments, but some may well be adapted with good results under a doctor's care, as I have done. They are planned to take off excess weight rapidly and benefit your health accordingly.

2. Most quick reducing diets limit your choice and intake of foods and therefore it is desirable that you take a general once-daily vitamin and mineral tablet. If you're over 40, take a once-daily 'therapeutic' vitamin tablet. Calcium pills may also be taken if a calcium deficiency exists.

3. These diets are for a limited time, to take off the overweight which most persons cannot lose by 'balanced dieting'. After most of the excess weight is off, you start counting calories and go on Stay Slim Eating of foods of your choice totalling 1,300–2,300 calories daily for women and 1,700–2,500 for men (see Chapter 7). However, I've had patients on the Quick Loss, fat-free, and other diets for a year or more. Vegetable and fruit diets can be continued indefinitely.

4. Since a quick reducing diet is, in effect, a deficiency routine, it is best to be checked by a doctor at the start and again periodically, particularly if you feel any ill effects whatsoever from the dieting.

5. You must stick to your restricted diet rigidly. No deviations or extras are permitted. By being faithful to your diet, your weight loss is swift, sure, and steady. Drink a great deal of water and as much coffee and tea without cream, milk, or sugar, also as large a quantity of non-caloric carbonated beverages as you wish.

6. As explained in detail earlier, in respect to the Quick Loss

method, when you reach your ideal weight and attractive figure, combine Stay Slim Eating with any of these quick action diets. Thus they can serve you not only in taking off excess weight rapidly right now but also in staying fit for the rest of your life.

9. Clearing Away False Fears About Quick Reducing Diets

Many persons have been blocked from dieting successfully by a fear that quick action dieting results in sudden weight loss that may be harmful. Another common concern is that the drop in weight may be followed by a gain when the diet is over. So they wail, 'What's the use?' Furthermore they're afraid that such dieting may result in a seesaw pattern of successive weight losses and gains which in their minds is injurious.

The fear that this may be bad for an individual's health has been fostered by many nutritionists, doctors, and other professionals, as well as by health faddists and others, who say that 'balanced eating' is the only right way to lose weight. *I have never seen or been given any proof that this is true – although I have openly challenged this concept for years.*

I have accumulated clear proof in successfully reducing thousands of men and women that 'balanced dieting' doesn't work with many people who are most in need of taking off dangerous, crushing excess weight. I have found without exception that with the patient who has no medical disorder the quick reducing diets I employ are safe and also are remarkably effective in a high percentage of cases. Nevertheless, certain prejudices held by doctors and the public die hard. Millions of overweights die earlier because of this.

As only one example of such prejudice, the accepted medical procedure after childbirth not long ago was to keep the mother in bed in the hospital for fourteen to eighteen days. Many doctors warned that cutting short this rest would result in impairment of future health for many new mothers. Today the new mother is permitted to walk about from a few hours to a day after delivery and to go home in a few days. Far from resulting in catastrophes, this change has speeded recovery to healthful, desirable activity.

Is restricted eating harmful or otherwise? Consider the case of US Senator Margaret Chase Smith. A newspaper report stated that 'Mrs Smith looks less than her sixty-six years. Trim as a model ... she usually starts her day at 6 AM and seldom gets home before 8.30 PM. Does she stuff herself with a variety of foods to produce the necessary energy? No. She lunches on powdered milk and cereal at her desk.'

Two Miss Americas have stated that they kept slim, healthy, and vigorous with very limited diets. When she became Miss America, Jacqueline Mayer said in an interview that she weighed 150 pounds at fifteen. She took off the excess weight by skipping lunch, going without breakfast, and not stuffing herself at dinner. She radiated health and beauty on this bizarre eating routine.

No diet here is a fraction as severe as that endured by Helen Klaben and Ralph Flores when their plane crashed in the Yukon wilderness in 1963. In the first week they finished their 'unbalanced diet' of tuna fish, sardines, canned fruit salad, and other odds and ends that usually would be consumed in a few days. Then they had 'only toothpaste and melted snow as nourishment'.

Aside from the lack of food, they endured terrible hardships, temperatures to less than 45 degrees below zero, frostbite, gangrene, a broken jaw, broken ribs. They had no special protective equipment, no blankets or bedding. Yet neither one apparently suffered any lasting organic damage from their forced fasting during the next forty-two days.

I certainly don't recommend that awful experience in order to lose weight. But I've pointed out this example to many people who told me fearfully, 'I'm afraid I'll damage my health unless I eat a "balanced diet", even if I follow a quick reducing diet for only short periods.' You'll learn otherwise, just as the airplane victims did.

In another case, one with which I was closely concerned, a young man of nineteen left for service in Africa where he immediately had to change his eating habits and cut down his food intake sharply. When he was checked by his family doctor before going, his health was noted as only fair, with some rather serious reservations.

During his year working for a medical service in the primitive country, his diet judged by American nutritional standards could hardly have been more unbalanced and inadequate. He could rarely get any meat, green vegetables, fruit, milk, eggs, or many of the common foods available at home and in his university dining room. Like the people of the country, he lived mostly on dishes made with 'mealies', tough Indian corn which is the staple food there, but is fed to cattle and pigs here. Very often he had to skip meals completely because there wasn't enough food around. In the remote mountains he went without sufficient food for weeks, gorging occasionally on meat when an animal was killed. In effect he lived on a semi-starvation diet.

He returned after the year without having had any illness while away. When his doctor examined him again, he found the boy a few pounds underweight but in better physical condition than when he had departed. The doctor commented, 'Maybe I should put many patients on semi-starvation diets with hard labour for a year.'

Compare the good sense of Quick Loss dieting with the operation of a good machine. Any mechanic will tell you that if you run the average machine at a normal working pace (ideal weight) day after day, you have the best chance of getting maximum service from it, without breakdowns. This kind of moderate operating pattern will usually insure longest working life and greatest efficiency from the machine.

When the machine is run hard day after day around the clock without any rest – overloaded, overworked continuously – that's when parts start to break down and fail. That constant, crushing overload week after week, month after month, is what causes the machine to break down completely before it would normally. That's when the machine *expires* before its time.

The same is true in general for the machine that is your body. If you're overweight, your machine is overloaded. Your parts (organs) are overworked, overburdened, day after day without a let-up. If you go on the Quick Loss Diet and take off 10 or 20 or more excess pounds in a few weeks, *you're quickly removing some of the load from your machine, relieving your heart and other organs.*

You don't burden your heart or the entire machine by taking away some of the fat load. By lightening the load, you provide some relief and rest. Even if you gain back some of the weight after a few weeks, your machine still had some relief for that lighter-load period. When you diet again and lose weight again, the machine gets further relief and is better able to function efficiently even if you add weight again later. With this see-saw pattern, your machine is getting rest and relief every time the load drops off.

Isn't it just good common sense? The more excess weight you carry, the harder your heart, other organs, your entire machine (system) must work to carry that burden. When you take off some of the fat, your machine has less work to do. It gets some relief, even if only temporary. This helps the machine until you eventually get down to your ideal weight or desirable operating load. Then your machine is able to work most smoothly and efficiently to achieve its longest operating life.

Are You Chained by Fear to an Iron Ball?

Consider another parallel to counteract false fears about see-saw reducing or quick losses of weight through dieting. Picture yourself carrying a twenty-pound iron ball in your arms for a block. You carry it a second block, a third, a fourth. You either collapse or feel like falling down because the burden is so great and seems so increasingly heavy.

Or – you carry the twenty-pound weight for a block. At that point the weight is taken out of your arms and immediately a ten-pound iron ball is substituted. You carry it for the second block. At the start of the third block, the lighter load is replaced by the twenty-pound weight. For the fourth block, you change to carrying a ten-pound weight.

What has happened? Obviously you're much fresher and better off with the seesaw exchange of the ten and twenty-pound weights. Every other block, your arms and body and your entire 'machine' enjoyed the blessed relief of having your burden cut down by 10 pounds. Your heart didn't have to work harder when it had 10 pounds less to support. On the contrary, it received relief, even though temporary.

By seesawing the weights over the four blocks, your machine had to carry and support an average excess weight of fifteen pounds instead of a steady burden of twenty pounds. Your body gets the same benefit of carrying a much lower average weight even if it takes you a year to remove that excess fat completely and even if you do it by seesaw reducing.

Certainly it's preferable to take off excess weight swiftly and steadily over a shorter period of time without increases after the initial sharp drop in weight. The primary benefit is that you reach ideal weight faster and thus relieve your 'machine' sooner without causing it to accelerate again when weight is added. But I hope you're convinced that if your weight should seesaw because you can't resist going back to over-eating after a stringent diet, you need not fear that this will harm your system.

As you'll see in the following chart, even if you lost 20 pounds and then regained it all over a period of four months, you still averaged 10 pounds lower in weight during that period. (The chart is not typical since very few of my patients went right back to their previous excess weight, but here you see what happens even if you did.)

As the chart shows, if this happened to you, that's still 10 pounds less on average that your heart had to drag around for the period. Thus your heart would get some relief and you would feel more health and energy even if only for the limited four months' time. Of course your goal should be steady loss of weight without sharp upswings.

)

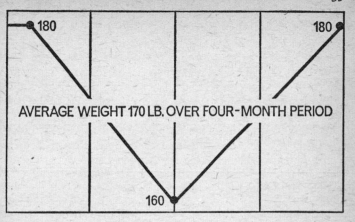

AVERAGE WEIGHT 170 LB. OVER FOUR-MONTH PERIOD

False Fear: 'Won't cutting food intake drastically hurt me?'

A fear expressed by many overweights is that they'll be harmed internally by cutting down food intake severely. I could fill a book with examples showing how wrong this concept is.

Even on semi-starvation and a close to starvation regimen during wars, few prisoners and others developed heart disease and such impairments from lack of food. Yet they suffered not only from starvation portions of poor quality foods, they also had to endure miserable living conditions, lack of sanitation, severe beatings, relentless interrogation, physical and mental tortures and worse.

The eating of the general populations in most European countries dropped drastically during World War II. Many people lacked proteins, vitamins, meats, and fats. Instead of growing weak and sick, most adults became healthier. They often lost their hypertension, shortness of breath, gall-bladder disease, and many other debilitating disorders.

Hospital statistics in many countries – Holland and Norway for example – showed a 40 per cent decrease in heart disease during the war, along with the sharp drop in food consumption and resultant general weight reduction. After the war ended and food became more plentiful, many people went back to their old habits of overeating and thus piling on excess fat. The percentages of those with heart disease and the other ills mentioned rose along with the increase in food consumption. By 1949 the tables on high incidence of coronary diseases and serious physical disorders were right back up where they had been in 1941.

Instead of worrying about being harmed by cutting down sharply on your food intake with a quick reducing diet, you'd better concentrate on the serious harm you're doing yourself if you're eating too much and carrying around that awful, debilitating burden of excess weight.

A doctor placed eleven obese patients on a fasting diet of nothing but water, vitamins, and minerals – no food at all. They fasted from twelve to one hundred and seventeen days. None of them could 'sneak in' any food as they were under constant watch in the hospital. All of them were in far better condition when reduced than when the fasting diets started.

That supervised fasting (definitely not recommended except under doctor's orders) was to help the very heavy person get down to normal or near normal as quickly as possible and then proceed to Stay Slim Eating.

False Fear: 'Won't unbalanced dieting and changing my eating habits hurt me?'

Many people fear that changing their eating habits or going on an 'unbalanced' diet will be harmful. They ignore the fact that their so-called 'balanced' eating habits produced the excess weight which ruins good looks and impairs health. Many have told me, 'I'm afraid that I'll be weak, dizzy, that I'll have "hunger headaches" and worse.'

When you are on any special diet, if you experience such undesirable symptoms and they don't disappear in a few days, your doctor should check your condition. In my experience, most of the people who complain of 'hunger headaches' and other such symptoms are the ones who have lost the least weight or none at all. In other words, they're experiencing ill effects even before the diet has really started or while cheating on the diet.

A popular story tells of a man complaining bitterly to a friend about his new diet. His friend asked, 'How long have you been on the diet?' The man answered, 'I start tomorrow.' This is not untypical of the complaining dieter.

The great majority of my patients who lose up to 10 pounds on the Quick Loss Diet the first week, and 3 to 7 pounds a week thereafter, usually comment 'No bad symptoms at all. No weakness. No headaches. Never felt better in my life.'

I've found that only a tiny percentage, from one to three of every hundred patients on quick reducing diets actually feel undesirable symptoms, a much lower percentage than on 'balanced' diets. Whether real or imagined, such reactions must be treated with understanding. Usually the 'symptoms' disappear after another day or two on the diet.

However, if an individual is mentally disturbed and his serenity and sanity are at stake with dieting, then obesity is to be preferred. This is not true of the overwhelming majority of overweights. For them a change in eating habits and an 'unbalanced', restricted diet is most healthful, not harmful.

Some people insist, like one patient: 'I'm naturally heavy and it's bad for me to be slim.' This is sheer self-deceit, unworthy of the intelligent person. In all my years of medical practice, I've never examined a normal person whom I've found to be healthier for being overweight than at ideal or average weight.

Another woman protested, 'Losing a lot of weight lowers my resistance and causes a lot more colds and other illnesses.' Patient after patient has used this as an excuse for staying fat. I keep emphasizing that getting excess fat off the stomach, pancreas, liver, heart, and other organs is most *helpful*. The entire system functions better. Colds and other ills are reduced rather than increased.

My experience with thousands of patients proves to me

conclusively that colds and other sickness are reduced as you reduce your weight. Your own good sense, plus the factual evidence, should prove to you that this is so. Does a truck motor run better when overloaded? Of course not. It wheezes and groans, parts break and finally the engine stops functioning. The same is true of the human overloaded with fat.

Another fallacy is the common statement, 'It's better to have a little extra fat on the body as a reserve for illness.' Actually every pound of overweight burdens the human system and impairs its functioning. The organs have the extra drag of feeding and carrying that 'extra layer' of fat. This impairs rather than helps the body in fighting illness.

If you elect to overeat and stay fat in spite of the hazards, that's up to you. Just don't delude yourself that you're healthier when fat. If you should ever decide to take off the extra blubber and get down to your ideal weight, your wonderful new sense of well-being and extra lightness and vigour will be your best proof that you're healthier when slim.

False Fear: 'Won't dieting bring on vitamin deficiency?'

This fear of dangerous vitamin deficiency is a common excuse for people afraid of dieting in general and quick loss dieting in particular. Even though millions in this country are deprived of enough food or the proper foods, there is very little evidence of actual vitamin deficiency.

Dr William Daniel Herbert, associate in medicine at the Harvard Medical School, stated, 'When all the cases of vitamin deficiency reported in the United States in a single year are added up, they do not reach the figure of 20,000, and therefore the clinically ascertainable incidence, as manifested by published reports, is much less than 1 per cent of the US population: 20,000 projected into a figure of 180,000,000. (Actually one-tenth of 1 per cent.)

There is no real chance of developing vitamin deficiency on the Quick Loss or other diets in this book. Don't be fooled by propagandists who imply that people can be 'vitamin-starved' even on three meals a day. Such vitamin deficiency is theoretical only. In spite of that, I suggest that while dieting you take any common once-daily vitamin and mineral product. On a fasting type of diet of only 300 to 400 calories daily, take a stronger 'therapeutic' vitamin product.

False Fear: 'Won't I look scrawny, haggard, and out of proportion if I lose weight quickly?'

Another common excuse for not taking off weight is the expressed fear, 'Even if I'm fat, I look strong and healthy now. I'm afraid to diet because I'll look weak and sickly. My skin will sag. The weight will come off in the wrong places and I'll look all out of proportion.'

This is an outright fake excuse. Down deep most of those who use it are aware of its falsity. All of us know too many overweight men and women who looked 'the picture of health' and suddenly took sick or died. It has been proved throughly by statistics that looking meaty and even 'pleasingly plump' is not a healthy look.

When you lose weight rapidly, your skin soon firms to normal proportions. About the only thing that sags on you are your oversized clothes. When people slim down they're invariably complimented on the radiant new look of fitness and beauty. It's an old-fashioned, outmoded idea that you should have 'meat on your bones'. The shocking facts about illness and earlier death among overweights, stated later, should dispel any lingering notion that a heavy look is a healthy look.

As for fat 'coming off in the wrong places', this is another fallacy. Overweight is overweight wherever it is located on your body. When you reduce, the weight comes off all over in time. If much of your weight is in your chest or neck, abdomen, pelvis, thighs, or calves, you must nevertheless keep going for total reduction of weight. The loss of weight may be greatest in areas where you're less concerned about losing it, but eventually weight loss is proportionate to your natural conformations.

If you reduce and keep your weight down, you'll eventually be rewarded by losing weight where you wish, as much as is possible. If your legs, for example, are naturally heavy, loss of weight will reduce them to a minimum but you can't expect them ever to be really thin. At ideal weight, you'll have your most pleasing proportions possible.

False Fear: 'I'm afraid I won't succeed'

I can promise you from my experience with thousands of different types of overweights that many who have failed before succeed with Quick Loss dieting as recommended here. As I've shown, even if you lose weight and gain some of it back, your body, heart, and other organs get the relief of comparative rest at least during the period when you were slimmer. You have everything to gain except pounds by trying my recommendations. Every minute of doubt you may have is a minute wasted.

To summarize, the big plus working for you to succeed with the Quick Weight Loss method, even if you've failed in dieting before, is the sharp loss of weight the first week. It proves quickly that you can succeed. Also helping you is the fact that your appetite decreases as the fat comes off. You break the habit of overeating. The overwhelming desire for rich food is usually gone. Your eating habits have changed for the better.

10. Additional Reducing Quick Tips And Special Aids

Over the years I've found that a number of simple 'quick tips' have helped former overweights to keep from letting excess fat creep up on them again. Some of these aids are noted elsewhere in this book, but bear brief repetition as helpful reminders and additional methods of staying slim.

More activity, less sleep. You have a better chance to become slim and stay at ideal weight if you limit your sleeping to a maximum of eight hours out of every twenty-four. There are several reasons. Activity, alertness, keeping on the go, all make reducing easier, as you use more energy and calories. You're less likely to eat unnecessarily because of listlessness and boredom.

When you're sleeping, you don't assist your body as much in burning up the food you've eaten. You usually use up to three times as many calories when awake and on the move than when asleep, as shown by the following averaged figures. While not universally accurate, these figures are helpful as a guide to general minimal utilization of calories.

MEN:

8 hours work, mostly standing ..	1,200 cals
1 hour activity getting ready for work	180
6 hours activity after work, walking, sitting, etc	720
1 hour activity preparatory to retiring	180
8 hours bed rest (sleep) (average)	540
Total calories used ..	2,820

WOMEN:

8 hours household activities ..	880
8 hours walking, sitting, resting ..	720
8 hours bed rest (sleep)	480
Total calories used ..	2,080

Note that many more calories per hour are burned up when awake and active than when asleep or resting abed. In effect, you deposit calories in your personal bank with the food you eat. You withdraw the calories by using up energy at work and play. If you deposit more calories than you withdraw, the fat deposits increase and your weight goes up. It's as simple as that.

The longer you sleep beyond eight hours daily, the more likely that you'll put on excess weight as you fail to use up calories from the food you've eaten during the day. The more active you are – working, walking, swimming, bicycling, bowling, golfing, and other pursuits – the more food calories you use up and the fewer will be left to settle in fat depots in your body.

Of course if you eat more and take in 200 calories of food for each 100 calories of energy expenditure, you'll keep your excess weight and increase it no matter how active you are. Even if you use up 3,500 calories daily, if you take in 4,500 calories of food you're depositing 1,000 excess calories that day to add fat to your body.

Eat slowly and you'll be satisfied with less food. Cut your food into many small portions, then take small mouthfuls and chew slowly and thoroughly before swallowing. Don't gulp liquids either, drink them slowly, s-t-r-e-t-c-h out a glass of skim milk or any other beverage or food – it will seem like more. Don't be a food bolter.

If you're having a banana and cottage cheese for lunch, for example, don't just peel the banana and eat it whole. Instead, cut the banana into many small slices and ring them around the cottage cheese on lettuce leaves. Eating slowly, you now have the feeling of eating a more substantial meal. Do the same with any fruit, even as an interval snack.

Don't eat or reach for food unless you're hungry. Remember how many times you've been disgusted with yourself when you've eaten something you didn't have to, didn't really want to. A neighbour stops in and you have a muffin (140 cals) with your coffee that you hadn't intended to eat. Or at the office you stop just for coffee and find that you've swallowed a doughnut (135 cals) without being at all hungry.

Before you reach for any food ask yourself whether you're really hungry. *If not, don't eat it.* Eating only when hungry can result in the elimination of many hundreds of calories daily. At meals, cease eating as soon as your hunger stops, don't keep eating just for the sake of eating or to finish the food on your plate.

Less food on your plate. It has been my personal observation, and the experience of patients, that if you put less food on your plate, you're less tempted to eat more than is necessary to soothe your hunger. The reverse is also true. If you put more cottage cheese or meat or other food on your plate than you really could do with, you're likely to finish the excess anyhow. It may amount to only one more meatball or just a tablespoon more of cottage cheese or an extra slice of bread, but it all adds up to extra calories you didn't need and wouldn't have consumed otherwise. It's practically a rule – the less on the plate, the less is eaten.

Before each meal when you're dieting, drink a glass or cup or more of water, tea, coffee, or non-caloric beverages. Just this little trick of pouring in the liquid *first*, is amazingly helpful in giving you a sense of fullness and in decreasing hunger pangs so that you're satisfied to eat less. You may finish the meal with a non-caloric beverage, and drink during the meal also. Every swallow helps fill you up without adding calories.

Omit salt or use it sparingly. Salted foods tend to retain more water in the body and increase weight. Unsalted foods are less tempting and you eat less.

Don't use sugar in beverages. Every teaspoonful adds 18 calories. If you ordinarily use 2 teaspoonsful of sugar in coffee, for instance, and drink 4 cupsful a day, you're adding 144 calories of sugar alone. Non-caloric sugar substitutes now are just as sweetening. You get plenty of carbohydrates and energy from the other foods you usually eat during the day.

Use salads abundantly when permitted on your particular diet (not permitted on the Quick Weight Loss Diet, for example). Ten large leaves of lettuce are only 25 calories. Using them as a tossed salad with non-caloric dressing makes a very filling and satisfying food as a snack or part of a meal.

Never use butter, margarine, oil, creams, fats, mayonnaise, and rich dressings unless prescribed on your diet. It you use 2 pats of butter at each meal (as too many overweights do), that totals 300 calories a day. Two tablespoons of mayonnaise on a salad adds 220 calories. Two tablespoons of cream in a cup of coffee, 4 cups a day, adds 280 calories (with 2 teaspoons of sugar per cup that totals 424 calories with coffee which alone has no calories at all).

Small calorie savings save pounds. You can lose 5–10–15–20 pounds in one year by eating the same foods as usual but cutting down on two or three items which you certainly won't miss. If you cut 50 calories a day, you save 18,250 calories in a year. There are about 3,500 calories to each pound, amounting to a saving of 5.2 pounds in a year.

To cut out 50 calories daily, eliminate 3 teaspoons of sugar (use non-caloric sweetening instead), or ¾ slice of bread, or 1 level teaspoon of butter. Thus you save 5.2 pounds a year very easily.

You can take 100 calories off your daily intake by cutting out 1 slice of bread and 2 teaspoons of sugar, or 1 straight alcoholic drink, or 1 glass of whole milk, or 1 glass of beer, or 1 glass of sugary soda (drink non-caloric instead). This slight change means 10.4 pounds less in the year.

Want to try for 150 calories saved per day? Eliminate 1 slice of bread and butter, or 1 martini or highball, or 1 tablespoon of rich salad dressing (use low-calorie dressing instead). Now you've saved 15.6 pounds during the year by this cutdown alone.

For those who put on weight slowly, hardly realizing it, this cutdown system is a simple form of natural reducing help in your Stay Slim Eating. However you must not replace the eliminated items with other indulgences.

If you now do a little more walking, golfing, bicycling, or other such exercise daily, you may save another 5 pounds over the year.

Avoid concentrated rich foods. Watch out for those high-calorie concentrated foods in the listing and skip them. If you pass up 4 tablespoons of hollandaise sauce on asparagus, for instance, you eliminate 200 calories (6 stalks of asparagus without butter or any sauce are only about 25 calories). Look down the calorie list and familiarize yourself with high-calorie concentrated foods to avoid, such as pork sausage (4 oz. – 340 cals), baked beans (1 cup – 320), candied sweet potatoes (medium – 295,) dates (pitted, 1 cup – 505), chocolate layer cake (2 ins. sector – 420), peanuts (shelled, $\frac{1}{2}$ cup – 420) and others. Surely it isn't worth using up to 20 per cent or so of your day's caloric allowance with a handful of peanuts or another unnecessary high-calorie food.

Taste the forbidden food but don't eat it. There's no need to deny yourself a taste of something you find particularly delicious, but remember that a taste has only a fraction of the calories of a portion. For instance at a party you might sample 1 small hors d'oeuvre to compliment your hostess, but if you eat one after another every time the trays are passed, you can easily consume over 1,000 calories *even before dinner starts*.

Break up your meal, save some for later. Remember the previous advice to stop eating when your hunger stops. Put aside whatever you haven't eaten for a snack later or for the next day. Many of my patients tell me that this one tip has saved them hundreds of calories daily.

Speak up: 'I'm dieting.' Don't hesitate to tell your hostess or people you're lunching or dining with that you're on a diet and must skip certain foods and have small portions of others. You can admire a dessert and compliment your hostess on how delicious it looks, but tell her firmly, 'I'm on a diet for my health and I can't afford to cheat. I must pass up the beautiful -dessert.'

You can explain the same way to business associates at lunch when you skip a cocktail. Chances are they'll admire your fortitude and wish they could emulate it.

It's not much of a hostess or host who feels 'insulted' because you pass up calorie-rich offerings or take smaller portions. After all, something very important is at stake – *your health and longer life*. It's far better to speak up than to be weighed down by dangerous fat.

Use low-calorie substitutes. One of your greatest aids to reducing and staying slim is to use the excellent low-calorie food substitutes that are fairly new and improving all the time. More and more packages are available to you with the calories marked, like a combination of tomato soup, beef stew, and peaches, 'a full three-course meal, less than 300 calories'.

Tasty salad dressings are available with practically no calories, instead of over 100 calories per tablespoon. A sugar-free jelly has only 10 calories a serving, compared with 200 calories per serving of ice-cream. There are many more such calorie-saving opportunities.

With all the fine sugar substitutes there's no reason to use sugar at 18 calories per teaspoon, about 770 callories per day.

Non-caloric carbonated beverages are a great aid. You can drink bottle after bottle, be refreshed and absorb hardly a calorie. Save about 100 calories per 8 oz. glass compared with sugary soda.

When eating in restaurants: Today you can eat out in almost any restaurant and still keep your calorie intake low. Keep in mind the tips and calorie tables in this book and use your good judgement, pause and choose, don't just order the first thing on the menu or that comes to your mind.

For example, in Stay Slim Eating, skip rich appetizers and creamed soups. Instead you can start with shrimps with a little cocktail sauce, or clear soup, clam broth (not clam chowder) or consomme. For your main course, choose meat, poultry, or fish, not fried. Have it brought without gravy or butter sauce, no butter on any of the food, no creamed vegetables. If you have a salad, have them leave off rich dressing; instead add a little vinegar or fresh lime or lemon juice. No bread or rolls.

Skip the dessert or have at most a half grapefruit, other fresh fruit, or jelly. Have plenty of tea or coffee during and after the meal, without milk, cream, or sugar.

Choosing in this fashion is easy enough and you won't seem like a freak. When you stop feeling hungry, stop eating; have the leftover food taken away.

Whenever it's not impolite, don't linger at the table. Get up, go into another room or take a walk. Of course you're less likely to eat when there's no food in front of you.

If you have lunch at the same restaurant daily, you can have them keep low-calorie foods for your convenience. Above all, take special care with everything you order. Even if it's only a hamburger, tell them to grill it without butter or fats, and not to give you a butter-soaked bun. These seemingly small caloric savings can save you extra pounds each week.

No second helpings. The second helping is one of the greatest curses in eating at home or as a guest, and you must make it one of your strictest rules to avoid this. Tell your wife or mother or hostess courteously but firmly, 'No thanks, no second helpings, I'm on a strict diet, please help me keep it.'

I have been appalled, after hearing a wife say virtuously to her husband, 'Nothing for you for dinner but lean roast beef and sliced tomatoes,' to see her then pile on another slice of roast beef and another as soon as the man's plate was clean. Then she tells friends, 'I can't understand why Henry keeps gaining weight, I feed him only lean meats, vegetables, and salads.' Those second (and third and fourth) helpings will wreck almost any diet.

Limit system. Limit yourself ahead of time by what you put on your plate, particularly when you're having a snack. If you're having biscuits and non-caloric soda, put one or two small biscuits on your plate, *then put the rest away.* Out of sight is out of stomach. Invariably if you leave the biscuits or cheese or other tasty foods in front of you, there's a good chance you'll keep gobbling mouthfuls – often without even realizing that you're eating.

Chewing gum helps some. I find that many of my patients have skipped their usual calorie-filled snacks by chewing gum instead. It may help you, and you can even get artificially sweetened gum, although there are few enough calories in an ordinary piece. You might also try dietetic gumdrops made with sugar substitute.

Toffee snack? Some people are able to assuage a craving by having a small piece of toffee. That's fine if you stop with one small piece once or twice a day. But what may happen is that you don't stop with one piece. A friend told me about three overweight secretaries in his office who had read that a piece of toffee will kill hunger pangs. They kept a dish filled with toffees on a table in the office. Each night the dish was empty and the girls were fuller and fatter. Instead of substituting one toffee for other foods, they were *adding* a load of sugar calories to their regular calorie-rich meals.

Keep a calorie and weight diary. It's been a great help to many of my patients to carry a little pad or slip of paper and keep track of the number of calories they consume meal by meal. When you write it down on paper that way you're more likely to stay within your daily limit and stop eating when you've reached it.

Also write down what you weighed each morning upon arising. You're much encouraged on the Quick Loss Diet to see your weight in writing diminish rapidly each day. On the other hand, when you're on Stay Slim Eating, if your weight is going up day by day as you record it, you see 'the handwriting on the wall' and you cut down or go on the Quick Loss Diet again.

Zero-calorie items. Keep in mind these items which contain zero or few calories and can be used as often as you wish on most diets (where not contra-indicated):

Carbonated beverages, non-caloric
Clear soup, consomme, bouillon
Herbs and spices, horseradish
Lemon and lime juice
Soda water
Tea and coffee
Vinegar
Spices

Use low-calorie vegetables. These are excellent foods which you may use in quantity unless not permitted (as on the Quick Loss Diet). Don't use butter or fats on the vegetables or in their preparation. The lowest-calorie vegetables are in the 5 per cent list, a little higher calories in the 10 per cent list, and the highest – but still relatively low-calorie foods – in the 15 per cent – 20 per cent listing.

5 per cent

Asparagus	Cucumbers	Mushrooms	Radishes
Bean sprouts	Endive	Parsley	Sauerkraut
Broccoli	French	Pimentos	Spinach
Cabbage	beans	Peppers	Vegetable
Cauliflower	Kale	(green,	marrow
Celery	Lettuce	red)	Watercress

10 per cent

Bamboo	Brussels	Eggplant	Tomatoes
shoots	sprouts	Leeks	Turnips
Beetroot	Carrots	Shallots	

15–20 per cent

Artichokes	Parsnips	Peas	Pumpkin
Onions			

Selective dairy products. Skim milk, cottage cheese, and cultured buttermilk should be kept in mind as excellent reducing aids and helps in staying slim (unless not permitted on your diet). These foods are relatively low in calories, considering their nutritive elements. Every time you drink a glass of skim milk (90) instead of whole milk (165) you save 75 calories, the difference being primarily in fats.

At first you may miss the richer flavour, but then you will probably prefer skim to the 'rich and fatty' taste of whole milk when sampled again. (An interesting sidelight is provided by a noted animal breeder who states that the most prized veal at the finest restaurants is from his skim-milk-fed calves.)

Diet a meal at a time. One trick used by some of my dieters is to keep the calories down for one meal at a time and not worry about the rest. For instance, you concentrate on your low-calorie breakfast, not giving any thought to being deprived of rich foods at lunch. Then you stick to a low-calorie lunch, not worrying about breaking down at dinner. At dinner, you maintain your restricted diet and you're through the day. But if you asked yourself on awakening, 'How am I ever going to abstain through three meals, or through a whole day?' you might have broken your resolution at breakfast.

The alcoholic is advised similarly by Alcoholics Anonymous not to worry that he'll never be able to drink again all his life, but instead to concentrate on the resolution, 'I will not have a drink *today*.'

Picture yourself. When an overweight patient first comes to me I urge that she have a snapshot taken of herself (or himself), preferably in a bathing suit. Then as she reduces she can compare her figure in the mirror every few days with the one in the photo. With the Quick Loss Diet, the difference is often excitingly apparent after only a week. Carry a 'before' snapshot and let your vanity take over to keep you from going back to being the horrible example in the picture.

It has been said that more diets have begun on bathing beaches and in dress shops than in doctors' offices. Some overweights have also helped themselves while reducing by buying a dress or suit that's too tight. Then they find that they work harder to take off the pounds and inches in order to fit into the new clothing properly and attractively.

Reward yourself. You may find it helpful to follow the

'reward system' used successfully by some of my patients. One young woman said that every time she was tempted to stop at a snack bar or coffee shop for a snack, she'd conquer the temptation and put the money she would have spent into a 'reward envelope' in her purse. When the change turned into dollars she'd buy herself a new article of clothing that she felt otherwise would be an extravagance but which she wanted very much. Saving those dollars and spending them made good sense in respect to her improved health and appearance as she slimmed down.

A family in my care with four members who were all overweight went on a 'charity diet'. For a month they put aside all the money they'd saved on costly desserts, liquors, and other expensive and fatty foods. They then gave the kitty to their favourite charity. They continued on this charity diet because they found it rewarding to the spirit as well as to the figure.

Instead-of's. When you 'must' have a snack in your Stay Slim Eating, keep in mind this list of possible snack foods that are relatively low in calories and can be taken *instead of* the high-calorie food you'd reach for without thinking.

Apple – 70
Buttermilk – 90
Cantaloupe, half – 40
Carrot, raw – 20
Cottage cheese, 1 oz. – 30
Cucumber – 25
Grapefruit, half – 55
Jelly, sugar-free – 10

Orange – 65
Plum – 30
Radishes (4) – 12
Shrimps (4) – 40
Tomato juice – 50
Coffee, tea, non-caloric
   soda – 0

Compare this with drinking sugar-rich sodas and whole milk; eating cake, biscuits, ice-cream, nuts, pie, pudding. You'll save hundreds of calories per snack by substituting the 'instead-of' – for example, a half-grapefruit, 55, instead of chocolate layer cake, 365 calories – saving 310 calories in one snack.

Don't feed a cold. There's no advantage really in the old adage, 'Feed a cold and starve a fever.' If you stuff yourself with food, whether you have a cold or a fever or not, you're putting an extra burden on your organs rather than helping them. If you have a fever, there's no disadvantage in eating normally. It all depends on your hunger. Most often the appetite diminishes with a cold or fever or both. You have nothing to gain except discomfort if you force food when not hungry.

Start Quick Loss dieting at once. If you're overweight, realize that there's no 'special time' for starting. The best time is now. If you delay, you're already beginning to lose the battle. No diet can reduce you while you think about it, you must go on it. The sooner you start, the quicker you'll lose weight, the better you'll feel and look. Have no fear of withdrawal symptoms such as experienced by those who go off cigarettes, alcohol and drugs. With Quick Loss dieting the only thing you lose is the excessive craving for food, soon after you start.

Cheating test. On the Quick Loss Diet (also on high-fat and fasting diets), a simple test will show whether or not you're following instructions as you should. Get a powder called acetest at a chemist's. When a drop of urine is placed on the powder it will turn purple if you're dieting correctly. But if you're cheating, the drop on the powder will remain white. This test has helped some of my patients turn honest with their dieting and lose weight more quickly and steadily.

Carry your Quick Loss Diet and Calorie Chart. Keep an extra copy of the Quick Weight Loss Diet and the Calorie Count Listing with you. If you're on one of the other diets copy it on a sheet of paper to carry. Referring to the instructions every time you eat will help you diet exactly as you should and lose weight most quickly.

A dieting friend helps. You may find it helpful, as others have, to diet along with an overweight friend. You can compare weight losses and check on each other to stay in line.

I know of twin brothers who were both about 30 pounds overweight. They went on diets the same day. At the end of each week, one twin would have to give the other a dollar for every pound he weighed more than his brother. When they were both at desired weight months later, each had won in better health and appearance.

Weigh yourself daily. As urged before, get on the bathroom scale each morning upon arising. If you see that 3-pound danger signal, go back on the Quick Loss Diet without delay.

Use seasonings to spice-up diets. A man complained to a friend, 'My wife is a confirmed weight-watcher. My problem is that the weight she's watching is *mine*!' If you're the cook of the family, one way you can help silence such complaints is to spice up low-calorie servings so that their flavour brings a smile instead of a frown. If you 'try a little harder', you'll find that it isn't too difficult to improve low-calorie foods from a taste viewpoint.

For instance, here are some suggestions from a very fine cook:

Boiled carrots, which are very low in calories, become much more delicious when you sprinkle them with grated or chopped, fresh mint leaves.

Use of monosodium glutamate products, now available in containers like salt, helps perk up the flavours of low-calorie vegetables, lean meats, and fish.

Chopped parsley and chives add a very tasty touch to plain boiled potatoes and other vegetables. You can grow them on any sunny window-sill. Clip and use the top inch and the plants grow right back again for further use and enjoyment.

Oregano and other flavourful Italian seasonings and spices sparkle up plain green beans and other good low-calorie vegetables.

Instead of using regular salt, which is better left out of most diets, make low-calorie foods more satisfying with no-sodium salt substitutes. Also, don't overlook the use of 'seasoned salt' made without sodium.

Juicy, fresh lemon wedges on the plate or table, a pretty little flagon of herb vinegar, other no-calorie aids, all help keep the dieter *on* the diet instead of slipping off.

There are dozens, even hundreds, of excellent, tasty relishes which improve low-calorie meals from a flavour viewpoint. These can be used in moderation at almost any meal without adding too many calories. Of course, good sense and judgement must be your guide. If you eat sweet pickle or many other relishes by the heaped tablespoonful (as I've seen some gluttons do), the calories will pile up. You'll be cheating yourself.

If a quick weight loss diet allows you the use of relish, that means a little relish for flavouring, not a full-sized portion like another vegetable. If you over-indulge in relishes, meal after meal, don't say, 'That diet doesn't work.' Instead, blame yourself for not working at your diet fairly and intelligently.

The same caution goes for all the flavouring tips in this book. Their purpose is to help the dieter enjoy his restricted portions thoroughly. If the result is to *increase* the portions or lead the dieter to grab second helpings, then it's better for his reducing success and good health to leave the foods bland and relatively flavourless. Nothing must be permitted to block the way to that one primary goal – *to get that weight off*.

Keep a refrigerator diet-snack container full. A lovely woman patient told me that she finally came to me determined to take off her excess 35 pounds when her husband kept calling her 'a *barrel* of fun'. She went on, 'I had been thinking that my friends regarded me as a happy-go-lucky companion. When my eyes were opened I realized that they considered me a fat clown.'

She stayed with the Quick Weight Loss Diet so faithfully that within a couple of months she slimmed down to become the most attractive wife in her social set. She told me that now, on her Stay Slim Eating, she finds that this is a help:

She took a short pretty glass vase and painted on it 'Stay-slim snacks'. She keeps this in her refrigerator filled with water (changed frequently) and a variety of carrot and celery sticks. Similarly, in small glass jars labelled the same way, she has handy low-calorie nibbles such as radishes and fresh, spiced (no oil) mushrooms.

All these snacks contain practically no calories. When she finds herself about to reach for a biscuit or sweet or other high-calorie snack, she heads for the refrigerator instead. Now her healthful impulse is to reach for a carrot stick instead of a chocolate-covered cherry. She emphasizes that the important point is to keep those 'stay-slim snacks' containers filled so that they're always at fingertip reach for instant snacking.

Guard against food and diet 'myths' that lead you astray. Many people have the craziest notions about what constitutes 'healthful eating'. Often the food myths have been fostered by constant, costly propaganda by special interests or misguided 'food and nutrition experts'. A patient and her overweight teenage daughter invariably failed in dieting because the mother believed firmly that everyone in the family had to start the day with a 'big, hot breakfast'. That's utter nonsense.

For the overweight, a breakfast of a chilled half-cantaloupe and artificially sweetened coffee or tea is far more healthful than the traditional bowl of steaming porridge, heaped with sugar and doused with whole milk or cream. It's infinitely better for you than a stack of buttered toast, or a platter of ham and eggs swimming in butter.

Even for the overweight youngster, a small ball of ice-cream and nothing else would be a more healthful breakfast since it would total fewer calories than those traditional 'hot breakfast' specialities. As for nutrition, there's as much or more desirable food value and energy in a bowl of high-protein cereal with skim milk, as in a similar serving of porridge with whole milk or cream.

It's all a matter of your own good sense as a guide. If you enjoy porridge, have it with skim milk or plain with artificial sweetening. That's better for you than if you add the high sugar and cream calories.

Another common myth is that artificially sweetened canned fruit and other foods are less nutritious than the same varieties in heavy sugar syrup. This simply is not true. You get enough 'energy' from other natural foods so that you don't need the sugar calories. By eating the artificially sweetened foods you lose only a load of fat-producing calories, not any vital nutritional elements. Once down to your ideal weight, your stay-slim eating can include sugar if you desire it, as long as you stay within the prescribed daily calorie intake.

If you continue to believe in the aforementioned and the many other misconceptions which lead you to add calories for the sake of nutrition, you're wilfully deluding yourself and harming your health and appearance. That goes for the whole family. If you're in doubt at any time, look up the scientific facts about a food, instead of being misled blindly by false clichés or advice of health faddists who frequently have some angle for personal gain.

Another misconception is that *Turkish baths* are a marvellous quick aid in reducing. This is false. A fat man may contend that he has stepped on a scale before and after a Turkish bath and seen an actual drop of several pounds. The explanation is that such loss is due to an outpouring of water, not fat. The water has been released in the process of perspiring heavily.

This weight loss is strictly temporary. The pounds go right back on in a few hours, as the scales would show if the same kind of comparison was made later. If you find a Turkish bath relaxing, if it makes you 'feel good', then enjoy it. But don't count on it as a dependable reducing aid. The same goes for body massage. There's no other way: you must change your eating habits meaningfully in order to enjoy quick and *continuing* weight loss.

Perhaps the most prevalent and disheartening myth of all,

I must stress again and again, is the terrible misconception that the only 'right way to reduce' is by losing 1 or 2 pounds a week, and that it's dangerous to lose a pound or more a day. I have before me a typical article from a national magazine read by more than ten million people monthly. It gives a 'sensible, balanced diet' and concludes that this 'brings a safe and satisfying loss of about 1½ pounds a week'.

So far as I'm concerned, based on over forty years of actual experience in reducing overweights, such statements are utterly *fraudulent* ... even though not intentionally so. The misguided overweight, especially the very heavy individual, will fail miserably on such a diet when he sees that he only loses 1 or 2 pounds a week.

He is being cheated when he is told that such a method will work for him. Yet, too many so-called nutritional experts, and even physicians without much experience of reducing patients in their practices, adhere obstinately and blindly to that outmoded viewpoint. Meanwhile those overweights fail again and again to reduce, and they moan, 'I'm one of those unfortunates who just can't get the fat off.'

You can see for yourself in one short week that there's no 'myth' about the Quick Weight Loss Diet. It will do the one, all-important thing that the 'sensible, balanced' diets fail to do – *it will get that weight off*. It will do this with the encouraging speed that is absolutely essential to reducing success for the vast majority of overweights. It will conclusively explode the other myth that you're one of those 'unfortunates' who can't get the fat off.

Realize what a difference quick weight loss makes from the standpoint of the dieter's morale, your own mental outlook: if you're an overweight man tipping the scale at 190 pounds, you'll lose 9 to 19 pounds the first week, if you're faithful to this diet which permits you a restricted variety of fine foods. If you're a heavy woman weighing 150 pounds, you'll lose 7 to 15 pounds the first week on this safe, pleasant method of eating.

The sooner you rid your mind of the misconception that

slow reducing is the only safe reducing, the quicker you'll take off the pounds. In no time at all you'll start seeing real encouragement in the mirror, as well as practically overnight on the scale. You'll be delighted to note the visible, sure signs that you'll soon reach the slim, attractive, youthful body buried under all the layers of unhealthy, burdensome fat.

Join a reducing club? Yes and no. Reducing clubs banding together a number of men and women, or both, with the common goal of losing weight, keep springing up in various communities. Such a club can accomplish lasting good in helping you to reduce if the members are guided by the right reducing methods. I've encouraged groups of individuals to diet together as a 'club' with regular meetings.

It's a real help when compatible people can get together weekly and compare the weight losses which are such heartening and *necessary* phases in arriving at the goal. As the members weigh in and are measured, they're ashamed if they haven't lost since the previous meeting. A lady gets a thrill of pride if the numbers show that she has done better than others.

An excellent system is to form a Quick Weight Loss Club which goes on the Quick Weight Loss Diet for the first few weeks. Then the members vote to switch together to the all-fruit diet for a week, back on the QWL Diet, then a week on the all-vegetable diet, back on QWL for a week, and so on. It's fun to arrive at a consensus of which quick reducing diet most members want each week after the first few on QWL. Later meetings help the individuals to maintain their ideal weight on Stay Slim Eating, with an occasional return to a week on the Quick Weight Loss Diet.

You must realize, as some overweights do not, that just joining a 'reducing club' is no guarantee that you'll lose pounds. If a club is based on 'slow, balanced dieting' principles, as too many are, plus a little group calisthenics, the rewards may be social, but that's all.

A discouraged husband, deeply upset about his formerly attractive wife's gross overweight, told me, 'She thought she'd solved everything by joining a reducing club for dieting and exercise. They ate more at the meetings or at the coffee sessions afterwards, than they would have if they'd stayed at home. All they really exercised were their mouths.'

He went on unhappily, 'They keep kidding themselves. My wife said that at the last meeting one member reported, "Yes, I lost weight last week. I went down from 150 to 149 9/10."'

Hopeless, after belonging to the misguided reducing club for two months, this same woman visited me at her husband's insistence. She lost 10 pounds her first week on the Quick Weight Loss Diet. She was excited and encouraged. She was on her way to saving her marriage. She could have done the same in the reducing club with the same effective method, but never on slow and unsure dieting.

Don't worry about food content, only calories. I learned in her first visit that a patient, who had never succeeded previously in dieting, was blocked primarily by an overabundance of empty knowledge about foods and nutrition. She considered herself an expert and immediately let me know it. She said, 'I know all about proper dieting so I don't think you can really help me. I understand full well that I must keep the carbohydrates down, the proteins up, the cholesterol down but the calcium up, the—'

I interrupted, 'The only important thing you have to know eventually is the calorie content of foods. You don't have to even know that until you go on Stay Slim Eating later. I'm going to start you on my Stillman Quick Weight Loss Diet where you don't even count calories, you just follow the simple, clear guides.'

It's too true that 'a little knowledge can be dangerous' in dieting. Another patient said, 'I'm worried about being restricted to the foods on the Quick Weight Loss Diet. That's because I've learned the body is a very complicated mechanism with a wide variety of chemical components as this shows.' She handed me a newspaper clipping containing this statement:

The body of a person weighing 140 pounds contains the following chemicals:

    enough fat for seven cakes of soap . . .

    enough carbon for 9,000 pencils . . .

    enough phosphorus to make 2,200 match heads . . .

    enough magnesium for a dose of salts . . .

    enough iron for a medium-size nail . . .

    enough lime for whitewashing a doghouse . . .

    enough sulphur to rid one dog of fleas . . .

    enough water to fill a gallon jug.

I returned the clipping and suggested to her, 'Paste this clipping in your scrapbook along with the snapshots that will keep reminding you how terribly overweight and misshapen you were. You stay with my diet and let me worry about the chemical composition of your body. That's my job.' Fortunately she put aside her fears and lost weight rapidly.

Beware of saying, 'I eat like a bird.' Whenever a patient tells me this, I point out that the problem here usually is definitely 'eating like a bird'. A bird does not generally eat much at a time. He pecks a little here and a little there. By the end of each day he has eaten quite a peck of food. That 'pecking' by humans all day or all evening or both can add up to an awful lot of calories.

At my suggestion, a worried wife kept track of an evening's accumulation of calories by her overweight husband. He complained constantly that he was gaining weight even though 'I eat like a bird, practically nothing all day long.' She kept a running listing of what he ate in a peck here and a peck there during the course of an evening of reading the newspapers, chatting a while, and watching television.

When she added up the calories 'per peck', the total came to over 2,000 calories! Her spouse didn't believe it until she showed him the listing and the figures in black and white. He later lost weight rapidly on the Quick Weight Loss Diet which doesn't permit such 'peckings' of high-calorie snacks.

Make a low-calorie checklist. There is a tendency for formerly overweight individuals on Stay Slim Eating to say now and then, 'I just don't know what to eat. Nothing appeals to me.' The problem is usually lazy thinking or a lack of imagination. You may find this little trick helpful:

Go through the pages of food listings in this book and put a checkmark alongside each food you like which is low in calories. Then, whenever you find yourself saying, 'I don't know what to eat,' open the book. Go through your checklist. You're sure to find plenty that appeals to you and will satisfy you most at that particular time.

Dieting with high blood pressure. One of my patients, a noted scientist in a field outside of medicine, was fearful of going on the high protein Quick Weight Loss Diet because he was 30 per cent overweight and had a high blood pressure of 180/100. From his 'scientific viewpoint', he figured that the burning up of his own body fat was equivalent to going on a high saturated fatty acid diet similar to eating butter, cream and animal fats.

Using some technical terms necessarily, the answer went like this: obesity and high blood pressure are very dangerous. Obesity without high blood pressure is dangerous. While it is so that the body will convert its own fat into cholesterol, it is also true that a low fat diet (exogenous) or your own body fat (endogenous), will not raise the cholesterol of the blood. The body can and does produce 150 to 170 milligrammes of cholesterol even if no fat is eaten. Also, the body maintains a blood sugar of 80 to 120 milligrammes whether little or no carbohydrates are eaten.

Experiments have shown that eating a great deal of fat produces excess cholesterol. Eating unsaturated fatty acids instead may reduce it. However, if you keep on eating large amounts of non-saturated oils, you still will have high cholesterol.

Overweight produces a fatty, 'creamy' blood. When you burn up your own fat, you may increase the amount of triglycerides in the blood, but this has not been shown to produce arteriosclerosis (thickening and hardening of the arteries). In reducing, it is more important as the primary goal to lose weight than to aim at reducing the cholesterol. Invariably the blood pressure will come down as your weight goes down to normal – and your life expectancy will be normal.

In summary, if you go on the QWL Diet, your cholesterol and the fatty acids in the blood will *not* be increased but will either remain normal or be lessened.

Is quick reducing desirable? This question keeps coming up. I have answered it various times in this book but re-emphasis may be necessary. My basic, *proved* answer is that among the thousands of my patients who have lost weight by quick-reducing methods, not one has developed any irreversible symptoms or diseases. All have looked better and felt better upon achieving ideal weight; this includes celebrated actors and actresses, musicians, composers, producers, doctors, scores of doctors' wives, many nurses, nutritionists, professors, and others.

Some patients brought up the fact that the Council on Foods and Nutrition of the American Medical Association has expressed the opinion that quick weight loss or large fluctuations may be harmful. I pointed out that if the Council believed this for certain, the ambiguous term 'may be' would not be used. Some doctors, 'nutritionists', and others have been crying 'wolf' for many years, but to date have not been able to show any harmful effects from sound quick reducing methods as recommended here.

Of course, a diet that has no vitamin C will produce scurvy eventually. If a diet has no vitamin B, it will produce beriberi. If it has no calcium, it will produce rickets or soft bones. If it has no protein, it will produce other types of disease. But this is assuming that it will be a *permanent* diet, not just for weeks or months but for years. Nowhere do I advocate such permanency.

With the Quick Weight Loss Diet and other quick reducing

diets, I have stated repeatedly that vitamin tablets or other available forms are to be taken daily to guard against deficiencies (which probably would not occur). In cases of gout or other diseases, I have stated emphatically that such dieting should be done under a physician's care. For the otherwise healthy woman or man, these quick reducing methods are essential lifesavers to get weight off and keep it off.

Use the QWL diet again and again. Please keep in mind from now on that I developed the Stillman Quick Weight Loss Diet as a method which will stand you in good stead whenever the occasion may arise that you must once again lose weight. You will have learned, like thousands of my patients, that it is very simple and easy to go back on the QWL Diet and *immediately* start losing the extra pounds.

Keeping your weight down permanently is a question of sustained willpower. The practical fact is that if a method of reducing takes too long, the average individual will soon lose hope, will be discouraged and will stay fat. You must be taught a method which is not only rapid but is easily repeated. This is it.

Day of deprivation. Many of my patients have found it surprisingly easy to set aside a 'day of deprivation' each week. On that day they eat no solid food, just drink quantities of water and other non-caloric liquids, and take a couple of vitamin tablets. By limiting the fasting to just one day they find it easier to stay slim. They say invariably that they feel better the next day. Instead of eating more the day after fasting they find they're eating less and losing weight more rapidly.

Stop at ideal weight. When you've reached your ideal weight, don't try to lose more pounds just to prove to yourself and others that you can. When I told a patient that she should indulge a little more rather than lose more weight, she answered, 'My dress size should be 12 but until this week I could only wear 18, then 16, and 14, and finally 12, now. But

I want to tell my friends I'm wearing a size 8!' I advised against this. It's healthier to be a little underweight than overweight. However, I recommend the happy medium – *ideal weight* – for feeling and looking your best.

11. Answers To Dieters' Common Excuses

You now have all the instructions you need for losing weight *successfully,* as so many in my care have done. I have found it has helped patients to give them in advance the answers to common excuses that I've heard innumerable times from overweights who are hesitant to diet or who are faltering. The facts are stated here in straight, common sense doctor-to-patient terms, as in my office. I hope you'll read them carefully, as they can help you achieve the healthful, attractive slimness you want.

> *'I heard that somebody died from dieting,*
> *so I'm going to stop my diet.'*

(One woman made this excuse after Quick Loss dieting had reduced her from 201 pounds to 146, and was headed for her ideal weight of 135.) In my entire experience no one in my care has ever become seriously ill from dieting, let alone worse than when overweight. When I've looked into other reported instances I've always found that dieting itself was never the cause. (She went on with her dieting and is now lovely and healthy at her desired weight, years later.)

> *'I felt fatigued so I ate and broke my diet.'*

In the great majority of cases a person doesn't eat because he's fatigued. He simply uses this as an excuse. I checked this patient thoroughly and said, 'There's no physical reason for you to feel tired all the time. Conditioning and environment have taught you to reach for something to eat without a reason, often even without hunger. Others will drink or smoke too much or chew their nails. You think that food in the stomach is the answer to tiredness so you use fatigue as an excuse to stuff yourself. If you'll fully recognize that it's an excuse and not a real bodily need you'll stop making the excuse and stop excess eating accordingly.' (He thought about it and went on dieting without any further 'feeling of fatigue'.)

I also tell patients who complain of fatigue to move around more and get more exercise – a very helpful antidote to that tired feeling overweights often get from characteristic inactivity. I suggest these rules: 'Don't lie down if you can sit. Don't sit if you can stand. Don't stand if you can walk.' All this minor activity not only wakes you up and helps dispel fatigue but it also increases your expenditure of calories.

> *'I sprained my ankle, felt sorry for myself,*
> *so I ate to appease myself.'*

I told her: 'Not only were you lame due to your ankle but you gave yourself a lame excuse. There's no physical relationship between a sprained ankle and a fatty stomach except that overweight probably contributed to your awkwardness and caused the accident. You'll be less likely to have such accidents when you become slim and graceful. You'll have better balance and reflexes. It's far better treatment for a sprained ankle to lose weight fast than to gain more.'

Another patient told me that she'd 'read somewhere' that drinking lots of rich milk – forbidden on her Quick Weight Loss Diet – would help her broken leg to knit. So she was gulping milk and had gained back ten of the pounds she had lost. She was very vague about where she'd found such incorrect information. When she faced the fact that she was deluding herself she went back on the diet and lost weight rapidly.

> *'I had to go off the diet because I craved biscuits.'*

Your answer to such an excuse is in the selection you have among over sixty diets in this book. You can't eat biscuits on the Quick Loss Diet because the carbohydrates break up the efficient functioning of the method, as explained. You can switch to the temptation diet, indulge your craving for biscuits and still lose weight, then return to the Quick Loss method and continue your rapid steady weight loss. I've given you a diet for 'cravers' – if necessary, use it.

'I ate when I wasn't supposed to because I had a "hunger headache".'

You'll find details about 'hunger headaches' elsewhere but the vital point here is that you're kidding yourself with such an excuse. If your headache is genuinely due to hunger you can satisfy it on the Quick Loss Diet by eating a hard-boiled egg or two tablespoons of cottage cheese or a chicken leg or a slice of lean meat or fish. 'Realize this,' I told this patient, 'and next time get rid of the headache, which probably won't even occur, by eating something that's on your diet instead of a forbidden calorie-rich food.'

'I lost 5 pounds on your Quick Loss Diet, then put it all back over a social weekend.'

Very few of my patients break their diet by the end of the first week and I blame myself for not getting my instructions across so that this woman would have the needed strong motivation. She finally realized that she was losing her attractiveness and endangering her health, and that saving herself was a matter of staying on the Quick Loss Diet every day for weeks, not breaking down fifty-two weekends of the year. She gained the firm realization of what she must do. She told her hostesses and guests that she was reducing as a matter of self-respect and life-or-death so far as she was concerned. By strengthening her resolve and pride she re-established the diet and went from 150 pounds to her desired weight of 115.

'I can't stick with my diet because of necessary heavy business luncheons.'

There are two ways to answer this excuse. Many of my businessmen patients have simply told clients and associates at lunch that, 'I'm slimming down for my health so I ask your indulgence. No reason for you not to eat what you want. In a few months I hope to be at my proper weight and in the best possible health. Then I won't have to be this strict.'

The other way, also effective, is to choose the one meal a day diet. Don't overeat at lunch but eat normally enough so that the matter of dieting doesn't come up. That's your meal for the day. Have coffee and vitamins for breakfast, coffee and cottage cheese for dinner.

> *'My menstrual period was different from usual,*
> *I became frightened and had to eat.'*

If something is wrong with your menses, as I told this patient, then eating certainly isn't going to cure it. Instead of gulping a big meal or a load of sweets, thus cloaking your symptoms by appetite gratification, you should see your doctor. He'll tell you what is really the trouble and will act to correct the problem. Swallowing an ice-cream soda isn't the solution. When something hurts you or worries you, get the proper treatment from a doctor – not from a dish of ice-cream. Breaking the diet won't solve a thing, it will only keep you from your goal of a slim, attractive, healthier body. (With this woman, her menstrual difficulty was quickly checked by administering a simple diuretic. She resumed Quick Loss dieting and stopped looking for relief from every problem by grabbing forbidden food. She has achieved her ideal weight and a beautiful figure, maintained for three years now.)

> *'I keep breaking my diet because my hostess piles food on my*
> *plate and I can't be ungracious so I eat it up.'*

There's no sin in leaving food on your plate as soon as you've eaten enough to stop your hunger. Put down your fork and don't pick it up again. 'The easiest way to get rid of weight is to leave it on your plate.' You're better off to stick with your diet and tell your hostess so *before* she piles up your plate. Anyone who would consider you 'ungracious' for being firm about improving your health and appearance isn't a worthy friend. Don't cloak your own weakness and sense of self-indulgence by calling it graciousness or politeness. Be firm next time and watch your hostess respect you for it.

'I get too irritable, depressed, and mean when dieting so I have to eat whatever my stomach craves.'

It's a weak excuse to blame irritability and depression on your stomach's cravings. You're evading the real reasons which may include idleness, unhappiness with your appearance and way of life. The fatter you get, the worse off you are. Get on the move, get interested in stimulating things as suggested on other pages. Treat your mind instead of your stomach. Success at becoming thin has improved the entire outlook of many of my patients.

'I can't go on a diet because I have ulcer-like pains in my stomach so I keep drinking milk all day and putting on weight.'

(I've heard this from many people who were not my patients but whom I've encountered socially.) If you have 'ulcer-like pains' then you either have an ulcer or think you have. You should see a doctor immediately instead of pouring in milk on top of your other food. The correct diet will counteract hyper-acidity in the stomach. If you are overweight, the doctor may very likely put you on a reducing diet which permits you six glasses of skim milk and six bananas a day and nothing else, alternated every other hour.

'My mother was sick so I ate to drown my troubles.'

This is one of the most common and sickest excuses of all, that, for one reason or another, 'I ate to drown my troubles.' Of course your emotional anxiety is very real. Overeating is one of the worst ways to treat it. Overeating when you're overweight especially drowns your heart and lungs, clogs your arteries, and shortens your life. If you wish to serve and help your mother (or solve other problems that produce emotional tension and anxiety) don't injure yourself by overeating. Then you may not be able to help because you've developed your own infirmities. Instead, direct your thoughts and energies into painting, writing, knitting, sewing, or other productive pursuits.

'I developed mouth ulcers from dieting so I had to eat.'

It's possible for allergic patients to develop mouth ulcers from foods, but upon questioning this patient further I learned that she had had mouth ulcers quite a few times *before* she started dieting. She had honestly had a complete lapse of memory about the previous times. Convinced that there was no relationship, she went back on her diet.

I can't begin to list all the ills and symptoms that patients attribute to dieting until they are reminded that they have experienced the same troubles long before. One woman claimed that she'd developed 'a lump in the throat' due to dieting. I diagnosed this as *globus* (ball) *hystericus* (emotional) which is a common symptom among nervous and tense individuals. She admitted that she'd had it in the past quite often. After her mind was relieved by understanding the cause, I told her to indulge in eating her favourite foods for a few days, then return to her Quick Loss Diet. Instead, her 'lump in the throat' vanished at once; she returned to dieting immediately and stuck with it. Once slimmed down she was much less nervous and tense.

'I was so bored that I ate and broke my diet.'

These cases are sad and all too common. I urge all such patients to become interested in some activity. In my practice, and in talking with people, I find that many heavy women eat because they haven't much else to occupy their time or mental and physical energies. There are exceptions, of course. By and large I've noted that older women with lessened responsibilities tend to be heaviest. Busy housewives occupied with children and community activities are more likely to be only slightly overweight. Alert career women and secretaries on the go are slimmer on the average. It helps you stay on a diet and then on effective Stay Slim Eating if you keep yourself busy, interested, active, on the move.

'*I had a sore throat so I took lots of honey and ice-cream to soothe it.*'

Don't demean yourself by using such an obviously weak excuse, treating your throat (or other ills) with doubtful or useless weight-producing 'remedies'. A sore throat needs bed rest and proper medication, not rich sweets that make your already overworked, fat-laden organs work harder than ever and pull down your overall health. Another patient told me that he had a sore tooth so he could only eat rich, creamy cereals, and ice-cream. I told him to see a dentist and to switch to the temptation diet. After a few days he called me apologetically and said he was back on the Quick Loss Diet and losing weight rapidly; he didn't fall off again.

'*I developed polio and was confined to bed and house for a long time without moving around so I gained weight terribly.*'

I told this person whom I met socially: 'It's too bad that you weren't instructed to eat very little. Since your expenditure of energy was so small, you needed very little food to supply your body's requirements. It happens to so many people, not ill, who cut down on their activity without cutting down on food intake. The only alternative to weight gain when there is a lack of activity and exercise is to eat less.' (The polio victim, with her strong will, went on the Quick Loss Diet and lost her excess weight rapidly. This was particularly important and helpful for her since overweight was further inhibiting her freedom of movement.)

'*I went to Europe, ate too much, and became very fat.*'

Can one really blame a trip to Europe for his own self-indulgence? Another of my overweight patients went to Europe and came back slimmer than when she left. She and her husband were absorbed by everything they saw and did. Their activity was increased by going to museums and places of interest, collecting antiques, working off a lot of energy (and calories). They ate in the famous restaurants, enjoying the food thoroughly without over-eating. They were aware that stuffing themselves could only add fat to their bodies and take away years of life. I urge you to slim down and watch your eating wherever you go so that you can stay healthy and enjoy a long, vigorous life.

> *'The skin on my neck was sagging so I tried to fill it out by eating.'*

Except for the very aged the natural elasticity of the skin will bring it back to normal soon when reducing, if it sags at all. If you're youthful this is no problem, so by all means reduce fast and early to avoid this temporary condition ever. In any case, at any age, be patient and your skin will return to normal. If you're quite elderly you'll feel in so much better health when slim that you won't mind wrinkling or sagging if it should occur.

> *'I've been a bad girl, I ate like mad.'*

There are many crimes that are much worse. You're not the first one to slip off a diet, especially if you're on a balanced, gradual diet. Start over again. This time pick the Quick Weight Loss Diet or another quick action diet that suits you best. Each time you go back on a diet (if you slip off) it becomes easier to stay on it and lose weight rapidly. Use your mental energy to stay on the diet instead of using it up in scolding yourself. By prodding yourself you can attain that slim figure which earns admiring (and envious) glances.

'After giving birth to my baby I wasn't able to take off the weight I gained. I couldn't stay on the diet.'

Especially if you were slim before having your baby (or at any weight), you can be just as slim afterwards unless your doctor has given you any reason otherwise. Any change affecting your weight is extremely rare. It's just as important to you and your happy married life (often even more essential) to stay slim and attractive. Once your body has recovered fully from giving birth, the dieting recommendations given here should work for you. This is true no matter how many times you may have given birth. If you overeat, don't blame your baby for your own unwarranted self-indulgence.

'I felt that I was getting anaemic and developing "tired blood" on the diet, so I quit.'

I advised this gentleman that he was developing 'tired blood' only in his overactive imagination. Blood tests proved conclusively that he was not 'getting anaemic'. Being a sensible businessman who believed the specific results of the tests, he went back to quick loss dieting. He phoned me a week later to say that instead of feeling tired he was enjoying a sense of extra vigour and buoyancy due to shedding excess, burdensome fat.

'Subconsciously I believe that I hate my mother, so I keep eating to stay fat and unattractive just to spite her.'

I didn't laugh at this because I hear that kind of thing very often from overweights. Such excuses are very popular today with self-psychoanalysts who grab at any reason to keep stuffing themselves. Usually such 'psychological anxiety' alibis are claptrap. Psychiatric problems should be treated by a psychiatrist, not by yourself, and certainly not by well-meaning but misled friends. I urged this patient to put aside thoughts of her relationship to her mother (at least as it influenced her eating) and to try the Quick Weight Loss Diet for one week. She was so pleased at losing nine pounds that week that nothing could have stopped her from continuing to reduce down to her ideal weight. In a later visit she told me, 'I love my figure and my mother.'

'I broke my diet with chocolate because suddenly I needed something sweet.'

Eating 'something sweet' today doesn't mean breaking your diet. A satisfying 'sweet' can be a glass of refreshing non-caloric beverage. You can also enjoy tea or black coffee with artificial sweetening. You can have handy some non-sugar chewing gum and artificially sweetened gumdrops to satisfy your sweet tooth without adding calories. However, don't gorge yourself on them.

'I ate because I had cramps in my legs from dieting, and heard that unless I ate a lot I would get a floating kidney, a hernia, or varicose veins.'

All this is utter nonsense yet I've heard such alibis again and again. I assure my patient that whoever told him this probably wanted him to stay heavy. The fact is that the skin and muscles tighten where the fat is lost. One of the best treatments for varicose veins is losing excess fat. It also helps relieve cramps in the legs and arthritis of the knee, ankle, and foot.

'Since I started taking "birth control pills", I've found myself overeating and putting on more weight.'

I convinced this patient that she'd only given birth to another excuse to indulge herself, since there was no valid reason why 'birth control pills' should cause her to overeat. (However, occasionally some women seem to retain a little extra fluid when first starting on the contraceptive pills.) Such medication contains no calories and does not act physiologically to increase the appetite. She protested weakly, 'A friend told me that these pills make a person eat more.' Once she realized that such pills have as little relation to the appetite as aspirin, she returned to her Stay Slim Eating, along with occasional weeks on quick loss dieting. She then kept her weight down and her figure lovely.

'I'm already 15 pounds overweight so I decided "what's the difference if I add a few more pounds" and I let myself go.'

I told this patient that each individual's system has its breaking point and that adding 'a few more pounds' or even one more pound may be just enough extra burden to bring on serious illness and disability. Every pound of excess fat taken off lessens such danger. She was persuaded, like many others, to think positively instead of negatively. Instead of letting 'a few more pounds' pile on, her daily goal – like a game – was transformed into seeing 'a few more pounds' vanish from the scale. On the Quick Weight Loss Diet she was able to see this positive result with her own eyes in the scale's 'numbers game' day after day. As her weight went down her spirits went up. She has become slender and attractive, and has been married. Whenever she sees 'a few more pounds' show on the scale she returns to the Quick Weight Loss Diet for a week or so.

'I break my diet at bedtime because I can't fall asleep unless I have a glass of milk and a sandwich or cakes and biscuits.'

This patient has convinced himself that only those fattening snacks could put him to sleep. He promised me that for the next week he'd substitute a moderate portion of cottage cheese and a non-caloric beverage as his bedtime treat. After three days he phoned to say that he'd enjoyed a full, sound sleep the night before. By the end of the week he'd stopped eating *anything* at bedtime and was sleeping well. Since then he has taken off most of the excess weight which was crowding his organs and preventing healthful, comfortable sleep. He's had no trouble with insomnia at ideal weight.

*'I was operated on for gallstones, put on a great deal of weight
and haven't been able to stick to a diet and take it off.'*

This woman was trying to reduce by cutting down on her eating gradually since she feared bad effects from a stringent diet.
She'd heard that it's healthiest to eat a lot after recuperating
from an operation, which is exactly wrong. I reassured her that
she'd recovered fully, and put her on the Quick Weight Loss
Diet. Within ten weeks she took off 30 pounds. She then went
on Stay Slim Eating for a month. She returned to quick loss
dieting and soon took off 10 more pounds which brought her
down to her ideal weight and figure. She realizes now that it
was excess fat which caused her weakness after recovering
from the operation, and that effective dieting brought back her
strength as well as attractive appearance.

*'My conscience troubles me if I leave good food on my plate
or throw it out, so I overeat.'*

I say to such patients, 'Please realize that one of the worst
things you can do to yourself in your overweight condition is
to stuff your system and add dangerous pounds. It becomes easy
to convince your "conscience" if you're once convinced of this
fact: the food you leave on your plate or throw out can't kill
you; overeating it can.'

*'I was so dry all week I kept gulping minerals
and up went my weight.'*

Reach for all the sweet minerals you want but next time make
sure that they're non-caloric. You can drink five or more refreshing bottles a day and add practically no calories instead
of over 500 calories daily from sugary drinks. There's no
difference in taste.

*'If I do anything wrong, I punish myself
by over-eating.'*

My answer to this woman was: 'The fact that you recognize

this common flimsy excuse as your own is encouraging. You're not the first to slip off a diet. But by over-eating, you're only doing something else wrong to your looks and your health, so you would have to punish yourself for over-eating by more over-eating. That's a silly, stupid cycle that you can certainly break. If you do anything wrong, or if anything goes wrong, you can cope with it better if you're healthier and more vigorous and alert than if you're burdened with fat and lethargic in mind and body. By following the Quick Loss Diet you'll be better equipped mentally and physically to cope with whatever goes wrong.'

*'I felt uncomfortable because I was constipated
so I ate a lot.'*

Constipation and lack of bowel movement can happen on any diet which doesn't contain much waste material. The Quick Loss method, also the formula, high fat and high protein, and fasting diets, for example, have very little residue so that a bowel movement is delayed. You may not have a movement for a week. Don't be concerned, because your worries are strictly psychological. A person can go up to four weeks without a bowel movement and suffer no ill effects. If you're concerned, take a mild laxative such as milk of magnesia or mineral oil.

'I can't afford to reduce because I have to look robust and prosperous in my business, not sickly and skinny like a dieter.'

I heard this excuse from a friend at lunch whom I'd warned about his alarmingly increasing overweight. Then he belched and grumbled, 'I'm full of gas. I have an important meeting at 2:30 and I'm worried about it.' I asked him, 'Do I look sickly and skinny?' He said, 'No, you're an exception.' I insisted, 'I'm not an exception. You're using an empty, fake excuse to justify your overindulgence and weakness of will. I want you to lose weight starting tomorrow, because I don't want to lose you before your time.'

'So I'm 40 pounds overweight but I'm the picture of health. I can lift my weight in wildcats and lick them too. No reason for me to reduce and become a skinny weakling. Besides, I've always been heavy and it hasn't hurt me.'

Sad to relate, this 'picture of health', in his early forties, was felled by a fatal heart attack one week after he told me this at a dinner party. Despite old-fashioned notions that still persist, weight gain leads to weakness, not strength. If you don't believe me, believe the death statistics and start getting down to your ideal weight right now. Whether or not you've 'always been heavy', overweight is dangerous and harmful. The sooner you realize this and start reducing, the better your chances for a healthy, longer life.

You've heard hundreds of other excuses like these and so have I. Each empty excuse is another way of saying that stuffing oneself with excessive, calorie-rich foods is more important than vigorous health, attractive appearance, and life itself. As far back as the 1600s, Gabriel Meurier put it this way:

'He who excuses himself accuses himself.'

12. Facts To Help 'Scare' The Fat Off You

I consider overweight the nation's No 1 health problem. It impairs and kills more people than any other sickness. I believe it is vital to publish these proved facts because, shocking as it is, most overweights must *diet or die younger*.

When Dr Louis M. Orr was president of the American Medical Association, he was asked in a newspaper interview, 'Do you consider cancer the greatest threat we face?' He answered, 'No. Cancer is the most dreaded disease in the United States. But the greatest danger to the health of the American people is obesity.'

If you are overweight and have any doubts about the importance and urgency of starting your Quick Loss dieting *now*, the facts here may convince you once and for all. The following death chart is based on a number of expert estimates. Some are higher, some lower, but these figures are generally agreed upon:

Death Chart from Overweight

If you are overweight, you will die younger, according to statistics. You will shorten your life by one to ten years or more, depending on how much excess fat you carry. It is conservatively estimated that 25 per cent of men in their thirties and 35 per cent of men in their fifties are about 20 pounds overweight. They are cut down before their life expectancy by heart disease, strokes, emphysema (shortness of breath, difficulty in breathing), diabetes, and many other afflictions. The following averages apply to men and women:

10 per cent overweight	10 per cent shorter life span
20 per cent overweight	20 per cent

30 per cent overweight	30 per cent shorter life span
40 per cent overweight	40 per cent "
50 per cent or more overweight	50 per cent to 200 per cent "

In other words, if you're 10 per cent overweight, your chances of dying earlier than your average life expectancy are 10 per cent higher. If your ideal weight is 160 pounds, and if your normal life expectancy is seventy, you're likely to die at sixty-three. You'll lose seven years of living. In effect you will have committed suicide at sixty-three. The more overweight you are, the more years you're wiping out.

Those are the frightening odds you're gambling with on your own life. Expert figures vary, as noted, but all agree that the more excess weight you carry, the more you invite earlier death.

If you're 10 per cent or more heavier than your ideal weight, then you should consider yourself an 'overweight' taking years off your life. Looking at it another way, you're making yourself look and feel years older than you really are. Your organs are ageing too fast for your years.

Even small increases in weight lower your life expectancy. If you're 5 per cent heavier than your ideal weight, you're not fat. But for health's sake, as well as improved appearance, you're much better off losing the few extra pounds and maintaining your ideal weight or slightly under.

Sickness in Overweights

The following is from a publication by the American Medical Association, warning about the dangers of overweight:

You're a fatty if you weigh 10 to 15 per cent more than your desirable weight. If you're supposed to weigh in around 130–140, and you're pushing 160, then you're one of the 34,000,000 Americans who are overweight.

What do a few extra pounds mean? It could be the

difference between good and bad health. In some cases, you're actually killing yourself by staying fat. Insurance companies can easily prove:

High blood pressure is found twice as frequently in fatties.

Hardening of the arteries is three times as much in the obese.

Diabetes and *arthritis* are somewhat more common diseases of those pleasingly plump.

The overweight are *poorer marginal risks*.

Serious heart disorders and *strokes* are more prevalent in the obese.

If you're not sufficiently impressed yet with the seriousness of overweight as a death hazard, consider also how it makes living far less pleasant. Even the famed French gourmet, Brillat-Savarin, warned: 'Well, go ahead and eat and grow fat, become ugly, heavy, have asthmatic attacks and die choked by your own fat.'

Overweights are more susceptible to many common and even minor diseases. They have less resistance to infection. They have more accidents than slim people. In my practice, fat persons were far more afflicted with worse and longer-lasting colds than those at ideal weight. They also complained of many more general vague 'miseries' which prevented or reduced the daily pleasures of living.

A medical newsletter states: 'Life Insurance companies have placed obesity high on the list of conditions which will considerably shorten life. The overweight person not only will not live as long as others, but he or she will also be a more frequent sufferer from fatigue, various bodily discomforts and malfunctions, and even serious and often crippling diseases such as diabetes, arteriosclerosis (hardening of the arteries), and certain cardiac conditions. (Obesity is the condition of being overweight when it reaches 10–20 per cent over normal weight.)'

Another medical report: 'Overweight produces a two and a

half times greater risk of coronary (heart) trouble than normal weight.'

Dermatologists warn that obesity promotes and accelerates a variety of skin disorders and accompanying undesirable symptoms.

Excess Fat Spurs Body Malformation

The top illustrations on the next page show how an ideal weight figure (left) is actually bent out of shape by excess weight (right), straining the muscles, body structure, and organs at many points.

How Overweight Spurs Foot, Leg, and Back Ills

The lower left illustration on page 184 shows how the weight of the body normally bisects the knee cap and falls between the large toe and second toe. Excessive overweight (lower right illustration) usually throws the feet outward. Weight does not fall at midline as desirable, but beyond the knee cap and towards the back of the arch. This may produce severe osteoarthritis (bone arthritis) of the pelvis, knee, and ankle joints, with severe flat feet, varicose veins, and general fatigue.

The Inside Story of Overweight Effect on Organs

The simplified drawings on page 186 indicate what happens in a man 5 ft 6 ins. tall, weighing 190 pounds instead of his ideal weight of 130–143.

The head and neck are enlarged due to fat deposits under the scalp skin and on the muscle layers. Fat deposits all over the body demand increases of blood pumped by the heart.

The heart is covered with a layer of fat, especially at the top. This is very burdensome to its proper functioning. The heart is pushed up into the neck and also brought forward. If you placed your fingers between the two collar bones, you would feel distinct pulsations, normally not apparent.

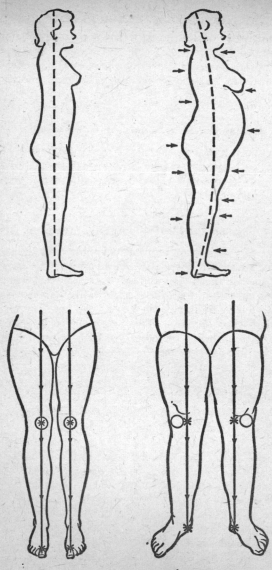

IDEAL WEIGHT   **OVERWEIGHT**

The main blood vessels are enlarged in order to carry blood to the excessive amount of fat deposited in the arms and chest wall.

Lung capacity is reduced at least one-third because of crowding by fat. The liver is enlarged to one and a half times its original size and is pushed up into the lung cage. The stomach is ballooned up to three times its normal size and pushes up into the chest. The abdomen is raised high because of the tremendous amount of fat in the abdominal cavity (a very large tumour or a large pregnancy acts similarly).

The coronary blood vessels cannot enlarge yet they must carry the same amount of blood to the heart muscle. If the rate of flow is increased by extra work or emotion or a huge meal, the heart cries out that the heart muscles are being undernourished because of scant supplies of available blood. In the picture at the right, note that there is plenty of breathing space and the heart does not reach up into the neck as at the left.

Dangers from Overstuffing with a Huge Meal

What happens when you're overweight and are overstuffed by a tremendous meal. The huge ballooning stomach starts pressing and irritating the dome of the diaphragm, often resulting in belching and hiccuping. The pressure is continued upward and impedes the action of the heart, moving it into a more horizontal position and making it more difficult for the coronary arteries to be filled with needed blood.

The upward pressures of the stomach and heart narrow the breathing space of the lungs so that the person's breathing is heavy and laboured. Liver and intestinal juices are poured out by the gallon. Eventually, with such excesses, the liver is likely to become fatty. The bile thickens. The pancreas may exhaust its power to produce not only juices for digestion, but also its supply of insulin.

As the excessively heavy meal is absorbed, the fat ingested reaches the blood vessels in such quantity that the blood

THE INSIDE STORY OF OVERWEIGHT EFFECT ON ORGANS

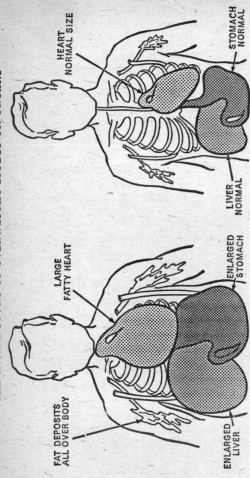

LARGE FATTY HEART

FAT DEPOSITS ALL OVER BODY

ENLARGED STOMACH

ENLARGED LIVER

MAN ABOUT 50 POUNDS OVERWEIGHT

HEART NORMAL SIZE

STOMACH NORMAL

LIVER NORMAL

SAME MAN AT ABOUT IDEAL WEIGHT

becomes thick and 'creamy'. Its motion is slowed. Then the blood may clot at the junction of the arteries, possibly producing either a coronary thrombosis or a cerebral thrombosis (coagulation). The fat may be deposited in the very small arteries. With repetition, the accumulations of fat may eventually clog and block the arteries.

If one eats too much over a long period, the liver which normally provides a clearing factor is damaged. The fat remains in the blood for a much longer time. This tends again to produced dreaded arteriosclerosis (hardening of the arteries).

Surgery Problems, Fatigue Afflict Overweights

Surgery is far more hazardous for those carrying excess layers of fat. Preparing a fat person for surgery, the anaethetist is more concerned. The surgeon must worry as he cuts through pounds and layers of fat. Even the wounds of the fat heal more slowly. Many times the surgeon will postpone surgery if the operation is not an emergency and send a fat patient back to the doctor to be reduced first.

Not the least of the common, everyday afflictions of most overweight individuals is chronic fatigue and a tired, draggedout, lethargic feeling which may interfere with normal sex life as well as other aspects of daily living.

If you need any further proof that overweight is a real and ever-present danger to the individual, realize that life insurance companies are cautious in issuing policies to fat people and often charge them higher rates. Such extra charges are then reduced to normal if the individual reduces and maintains his ideal weight. It even pays in dollars and cents here to slim down.

I've touched only superficially on the harm excess weight can do to the heart, blood vessels, and other organs. My purpose is to jolt you into starting to reduce now if you're overweight.

Your Ball and Chain of Fat

If you're 20 pounds overweight, for instance, you're dragging a terribly burdensome ball-and-chain of fat behind you with every step you take. That unrelenting weight is pulling on your heart and entire system. Even when you stand still, sit, or lie down, your heart and organs must support that twenty-pound ball-and-chain. To stop being a prisoner of that ever-present, debilitating load, start a successful quick reducing programme now.

Life or Death? It's up to You

Yes, you can mention a number of overweight people who live to seventy, eighty, and over, and who say, 'Being fat didn't keep me from living to a ripe old age.' Also, granted that there are some fatties who seem to be bubbling with energy – these are the rare exceptions. The facts prove that the odds are against long life or average life expectancy if you're overweight. Face it, as a whole, fat people die younger. Every time I see a grossly overweight person, I want to speak a warning: 'Get that weight off, you're heading for an earlier death. Reduce and save your life!'

Don't try to excuse yourself, as many do, by claiming that, 'most of my excess weight is muscle, I'm not really fat.' Overweight is overweight whether it's muscular or fatty. The fatty kind is more injurious but excess weight of any kind is a dangerous drag on a heart that has to pump extra blood and carry the extra load. Few over-muscled individuals are noted for longevity.

Many heavy patients have reported that afflictions they've suffered for years – backaches, headaches, gastric upsets, breathing difficulties, and other ills – have vanished along with excess weight. You'll enjoy many blessings when you've rid yourself of that extra load of ten or more pounds of burdensome fat. But don't consider yourself free of all ills automatically; slimming is not a cure-all. Have your doctor give you a complete check-up to make sure that you're in excellent condition all over.

Keep in mind always that calories count and can kill; but just counting calories won't help. Only reducing your calorie intake drastically and effectively will cause you to lose weight rapidly. The Quick Weight Loss Diet will take off that excess fat quickly and give you a new lease on a healthier life, a better chance for a happier life.

'How Dear is Life to You?'

'The answer to the question for the individual is that life is priceless. To the family its loss is a major tragedy.'

That statement in relation to the importance of proper health care was made by Dr Willard C. Rappleye, Dean Emeritus of the Faculty of Medicine, Columbia University. I suggest that you ask yourself right now, 'How dear is my life to my family and me?'

If you're overweight, particularly if you're considerably overweight, you're not holding life very dear. Even if life may not be priceless to you, realize that your loss would be a major tragedy to your family. Unless you get rid of that excess weight and slim down to your ideal weight, you're hastening your death. The preceding plain facts and many others prove it beyond question.

Unfortunately, the warnings and recommendations of preventive medicine are little heeded by most people. When a fat man or woman is told to reduce to lessen the danger of a heart attack or other serious damage, he or she usually waits until stricken and near death before taking the weight off.

Even then many victims start loading on the excess fat once more after they're up and about for a while. Perhaps the facts here will convince you to begin the Quick Weight Loss Diet or other effective reducing diets now.

There's grim humour in the statement by a commentator that 'Things tend to even up – the more bodily weight you carry around, the shorter time you'll likely have to carry it.' An old proverb tells that 'Over-eating destroys more than hunger.' It destroys health and life itself.

Centuries ago, Hippocrates warned in his 44th Aphorism, 'People who are fat are apt to die earlier than those who are slender.' Centuries later 'fat people' haven't learned to accept this, often because they think there's little hope of success in reducing. My recommendations here have helped others who had failed in their efforts before. Hopefully my methods will help you succeed where your previous attempts didn't.

On the other hand, if you never have decided and tried to take off burdensome excess weight, I trust that you'll start at once. Saying 'I'm only 10 per cent overweight, I'm not a fatty' is not a valid excuse. Once you start to let yourself go the danger increases of undesirable, unbecoming fat piling on.

Recently a beautiful young woman whom I had helped to become slim and who had stayed that way, told me how she had finally managed to get her husband to reduce. 'During our twelve years of married life,' she said, 'I couldn't get George to take off the fat that was turning a handsome man into an uglier one each year. He kept putting on weight until he scaled over 240 pounds instead of the normal 175 pounds he was when I married him. I tried to get him to visit you or at least go on your diet, but he refused. He always answered, "I'm a happy fat man." '

She shook her head. 'I almost gave up. Last summer we rented a house at the beach. Most of the time George wouldn't get into his swim trunks even though he loves to swim. The few times he did go swimming he looked like a tremendous beach ball with legs. No one would ever know that there was a very handsome, tall, thin man hiding inside that ugly mass of flab.

'It didn't help when I told him again and again that I was afraid of his premature death because of his shortness of breath and other evidences of deteriorating health. Finally I said straight out, "George, I'm ashamed of the way you look. I'm beginning to dislike being with you. Either you start slimming down or I'm afraid that I'll eventually get to the point where I'll have to leave you!" '

She beamed. 'It worked. He couldn't get upset about dying before his time. But this blow to his vanity, and even suggesting that his overweight might be repulsive to me, as well as to others, and could break up our marriage – that did the trick. He made a deal with me that he'd go on your Quick Weight Loss Diet for just one short week. If he didn't lose weight – he thought he was one of those people who couldn't – I promised I'd shut up about it from then on.

'He's a man of his word. He stayed with the diet. In that one short week he lost 17 pounds. Each morning he couldn't wait to step on the scale and shout to me, "Down three more pounds!"

'In two weeks he has dropped 26 pounds. He's starting to look young and handsome again. He's so thrilled by the change, and how easily and pleasantly it happens on your method, that nothing can stop him now!'

This case fortunately has a happy ending. Too many others I've known about have ended in family tragedy with a woman leaving a grossly overweight husband whom she couldn't stand to look at or live with any more. Even more often the situation has been that of a man leaving his overweight wife who had lost her figure, her looks, her pride and spirit and finally her husband.

It's up to You

When I see an obese individual shovelling in the rich food, my impulse is to snap at him, 'Stop, you're digging your grave with your fork!' Of course I don't speak out. It's not polite. It's not considered gracious in our civilized society to tell someone that he's committing suicide by overeating. It's not 'nice' to warn a well-meaning woman that she's pushing her husband into an early grave by over-feeding him.

As for the reactions of friends and acquaintances, they're

not going to tell you frankly that you're losing your looks and your sex appeal by piling on ungainly pounds. The only one that will tell you is your mirror, and then only if you look at the reflected image clearly and don't erase the unbecoming fat bulges with your eyes.

I tell you specifically on these pages how to diet effectively. The proof that the method can work wonderfully is written in the tens of thousands of unwanted pounds dropped by my patients. But neither I nor anyone else can diet for you. I'm hopeful that you'll start taking off weight determinedly now. Further, that every time you may be tempted to stop dieting and go back to putting on fat, you'll first re-read the gruesome facts in this chapter and that they'll help keep you on your diet.

Remind yourself:

'I don't want to suffer from sickness and pain . . .'

'I don't want to be stricken by a heart attack brought on by self-indulgent overweight . . .'

'I don't want to die before my time because of gluttonous overeating . . .'

'I don't want to lose my looks under a blanket of lard and become an unpleasant sight, a misfit, seeming years older than I am . . .'

The means are here to start taking off excess weight now. As in the case of George, give yourself just one short week on the Quick Weight Loss Diet. See what wonderful things can start to happen for your health, your good looks, and your whole future.

13. Important Tips On Dieting Psychology

'The only thing we have to fear is fear itself,' was the pivotal statement in the historic speech by President Franklin D. Roosevelt which started to turn the tide and pull Americans out of the doldrums of the great depression.

In my experience the principal psychological reason why so many overweights have failed to reduce is *fear* – fear of injury from dieting rather than fear of sickness and death from not dieting. The most damaging aspect of this ingrained deterrent is fear of losing weight too fast, of cutting down on food intake, of eating in a different manner.

Much of this fear is inculcated by the old-fashioned but still prevalent idea of mother scolding, 'Eat! Eat! It's good for you!' Almost all of us have been part of the scene of a child not hungry but mother exhorting, 'Make yourself eat, otherwise you'll get weak and sick.' The common wifely or motherly reaction, even today, 'What? You had nothing to eat all day yesterday but a little grilled meat and lettuce on a diet? Are you trying to kill yourself? You'll get weak, you'll fall apart – eat! eat! eat!'

It's part of the general psychology about dieting that an all too common reaction about quick action dieting is, 'I heard about that actress, what's-her-name, who went on a severe diet and died of starvation – and with all the money in the world and everything to live for. Terrible!' Or, 'Who was that famous actor, years ago, who had to get thin to play Hamlet and he got sick and died? Not for me!'

This is all obvious nonsense, but it has been spouted repeatedly by otherwise intelligent individuals who can then devour their fattening foods with false self-justification. Tracking down such cases invariably proves that the persons, if they existed at all, were mentally sick and died of a cause completely unrelated to dieting. Furthermore such individuals, especially the rich and famous, were under the care of physicians who would have stopped any harmful dieting immediately.

Weak Excuses for Personal Weakness

Clearly all these are just excuses not to diet, not to break up one's cherished and indulgent eating habits, not to deny oneself 'the good things in life', even at the risk of ill health and earlier death. I state repeatedly in this book that it's best to be checked by a doctor before dieting, and certainly at any sign that the diet may be harmful. The doctor's tests reveal quickly whether any undesirable symptoms are physical or psychological disturbances.

I state again that if the self-denial involved in dieting results in mental or other illness, it is preferable to stop dieting and stay fat. But I must reiterate conclusively that from my long and thorough experience my quick action diets and instructions as detailed here are safe, quick, effective life savers for the 95 per cent of people who have no metabolic or other serious disturbances. I have reduced thousands of men and women who had failed in many previous attempts, not only by giving them the effective quick loss dieting techniques, but also by helping them to get rid of the psychology of fear of quick weight loss.

Unfortunately some doctors and nutritionists contribute to and bolster the harmful psychological fears by insisting that the only 'proper' way to lose weight is through 'balanced dieting', 'normal dieting methods' of eating a little less and cutting out a few high-calorie foods. Apparently it's not significant to them that such advice doesn't work for the overweights who try and fail, those who keep getting heavier and sicker from carrying burdensome fat.

To replace your own possible psychological fear of quick action dieting with a true and sound fear of not dieting and staying dangerously overweight, I urge you to read the proof to the contrary on these pages again and again. *Eat scared!*

The Right Frame of Mind for Losing Weight

You must develop a firm almost fanatic desire and resolve to lose dangerous excess weight; that's the reason for giving you the 'Facts to help "scare" the fat off you' (Chapter 12), and 'Clearing away false fears about quick-reducing diets' (Chapter 9). You must think, breathe, act, and eat with the conviction that you'll achieve your target of reducing to healthful, ideal weight, and that you're getting nearer to the successful result every day.

It may be a little exasperating at first when you cut down on food you've become accustomed to like. You may feel the pangs of hunger for a few days. You may be tempted, but be sustained by the knowledge and certainty that *you're going to succeed this time*. You're going to realize sensibly that it's wasteful to worry about gaining back the weight you lose.

You must believe and keep reminding yourself constantly, that quick loss dieting techniques will cause you to get down to desired weight. Then you'll find that following effective Stay Slim Eating will help you stay thin from then on, with an assist from quick loss dieting as recommended whenever necessary. You're not just getting 'empty advice' here, you're being given *methods that work*, and can work for you.

Common Alibi: 'Emotional Frustration'

You accomplish nothing by continuing to blame your present obesity on emotional frustration. I advise you to go on the Quick Weight Loss Diet or other of the quick action diets of your choice right now. You'll prove to yourself by the first week's dramatic drop in weight that you can reduce. That proof is your greatest prop in continuing the diet. I know that in a few weeks you'll feel much better and look much better.

You'll find that you actually lose much of your desire for calorie-rich foods. You'll be delighted that you undertook this reducing programme, and encouraged by the clear results, you'll stay with it until eventually you're as slim as you wish and should be.

Will you be crushed by a feeling of failure if, after losing quite a few pounds, you slip and start gaining again? No, not if you read again at that time, and time after time, the advice in this book. Realize that up-and-down reducing is neither uncommon nor harmful. Don't expect to succeed in a day; even when you hammer in a nail you can't do it with one blow, you must pound it again and again until it's firmly fixed.

By returning to the quick loss diet again, you'll lose again. And each dieting period results in lessening your desire for calorie-heavy foods. At the same time the sizeable drop in poundage, accompanying personal gratification, and compliments from others bolster your will to become thin. With each successive quick loss diet it becomes easier for most people to lose weight and keep taking off that excess fat.

When dieting, the warm, understanding, persistent encouragement of your doctor and your family and friends is very helpful. A cartoon noted a despairing fat lady being told by a salesman of bathroom scales, 'As you step on it, it moans sympathetically.' The most important encouragement you get from quick loss dieting is the sharp drop in numbers on the scale week after week and the slimmer image reflected by your mirror.

This is why the Quick Weight Loss Diet and other diets here are so helpful psychologically, as well as physically. This indisputable evidence that the diet is working and that your goal of a slim figure is not too far away often makes the difference between success or failure in sticking with your dieting resolve.

Be Proud of Your Willpower

Dieting requires willpower on your part and you should be proud of yourself when you succeed. Cutting down on food, either gradually or suddenly (the latter is my usual recommendation), causes both physiological and psychological changes. If you're one of the minority who experiences 'hunger headaches', fatigue, or other undesirable symptoms, bear with them a few days and they'll usually go away. If they persist, see a doctor if it's only to have him tell you that there's nothing physically wrong. Usually the symptoms then vanish quickly and you're on the way to the sharp weight loss you want.

Be proud and be positive about your dieting. Don't act like a martyr or feel sorry for yourself. When you refuse desserts, second helpings, or rich hors d'oeuvres at a party and others murmur, 'You poor thing!' – speak up. Tell them that you're pleased to be dieting, that you're proud of the pounds you've lost to date and that you're determined to get right down to ideal weight. Every pound you take off and keep off is a small miracle worthy of self-congratulation.

Quote Jean Kerr's amusing line, 'I feel about diets the way I feel about aeroplanes. They are wonderful things for other people to go on.' Be proud that you're one of the 'other people' who is accomplishing what every overweight would secretly like to. You're far stronger than the fatties who sigh, 'I wish I could take off this weight, but I know that the only thing I'll lose will be the diet instructions.' You're way ahead of them – you've stopped wishing, you're acting! The time is past when it was a status symbol to be fat in order to show prosperity; today the slim people are the ones who are envied.

Don't Overrate the Value of Food

I've been amused by a cartoon showing two children impatiently waiting for food to be served at their mother's bridge party. One tells the other, 'We'll eat soon. When they start talking about their diets they all get hungry.' One thing that keeps many fat people that way is the psychological aspect of rating 'good food' too highly in life.

I suggest that you regard high-calorie foods as your enemies. When you look at a piece of layer cake or other fattening delicacies, don't fail to see that 'B-E-W-A-R-E' sign on it. If you find yourself weakening towards such temptations, re-read 'Facts to scare the fat off you'.

Keep these two points in mind. First, there are many things more important in life than stuffing yourself with the 'finest foods' in the 'finest restaurants', prepared by the 'finest cook in town'. There is appreciation of music, art, literature, your family, and mostly your health and your longer life. Buy yourself a good book, an absorbing magazine; treat yourself to a movie, theatre, or concert with the money you'd have spent on rich foods. Feed your mind instead of your stomach for deeper enjoyment and lasting fulfilment.

Second, you can still be a gourmet by sampling or taking small portions of fine foods and really taking the time, eating very slowly, to appreciate their special flavours. The time will come when you can increase to moderate portions after you've finished your stringent dieting and are on Stay Slim Eating. The gains are certainly worth the sacrifice for the limited time it will take you to bring down your weight on the Quick Weight Loss Diet, for instance.

Facing Basic Insecurity

A funny, but somehow sad, line quoted to me was a wife asking her husband, 'Be careful, this is a trick question. Did you love me more before I got fat?' This reflects the basic insecurity of most fat people who know underneath that they are less attractive physically and sexually, whether they admit it or not. Psychologically this is a corrosive influence, often souring and even spoiling many aspects of single and married life. Isn't it worth the sacrifice of denying yourself over-indulgence in food in order to attain greater peace of mind and happiness for yourself and your loved ones?

It may also help keep you dieting to realize that you'll probably be saving money as you save calories. This point is sometimes a help psychologically, even to those who are quite wealthy.

Remember my previous advice – don't look dismally at the start into your future of weeks and perhaps months of dieting. There's an ancient saying that 'a journey of a thousand miles begins with a single step'. Concentrate on the first step and the next, not on the entire thousand miles. Keep telling yourself each morning after a look at the encouraging scale (if you haven't been cheating), 'I will diet *one more day*.' Each day it becomes easier as your appetite diminishes with your decreasing weight.

Psychologically and calorie-wise it's good sense to keep busy at work, play, every possible interesting activity. Face it, you're breaking a pleasurable, entrenched habit when you deny yourself unlimited eating. Use every diversion you can. Many failures are due to idleness, boredom and nervousness which lead to the busy, compulsive habit of eating and swallowing, eating and swallowing.

Don't Kid Yourself When Dieting

Losing weight isn't a breeze or a lighthearted matter, nor is it 'kid stuff'. Your health and length of life are at stake. So don't kid yourself that you're adhering to a diet if you're not. It helps many dieters psychologically to write down every item they eat so they don't fool themselves into saying, 'I had practically nothing to eat today.' Remember that 'memory is the greatest liar of all'. When you make a little list, the record is solid and inescapable.

It's not uncommon for a fat lady to delude herself, 'I don't have to worry about my weight any more, I bought a new girdle.' Or for a man to buy clothing two sizes too large in order to feel that he's thin. Instead be tough on yourself like the determined matron who told her waiter, 'If I order dessert, say, "Nix, Fatso!" '

If you find yourself making alibis for stopping your dieting, re-read the common excuses that I've noted in Chapter 11. Hopefully the answers will prevent you from making such excuses for yourself and will start you and keep you dieting successfully.

14. Getting The Most From Low-Calorie Foods

Low-calorie foods developed in the past few years amount to a small revolution in food processing and a great boon for dieters. Many of these foods are wholesome, nutritious and sometimes just about as delicious as their high-calorie counterparts. More and more women are beginning to take advantage of such reducing aids while shopping in the supermarkets.

If you're not getting all their slimming benefits now, you should start at once to help fill out your menus with low-calorie versions of such foods as are available already. A supermarket checker complained that some women are beginning to ask him to add up the calories of the foods they buy, along with the prices. He told one, 'You figure out your own calories, ma'am. I can only tell you what the stuff costs!'

One big advantage of these 'low-calorie foods' is that you can count the calories right on the package as you buy and as you use them. This saves you the trouble of figuring out the calories for yourself from a chart or from the personal knowledge that you soon gain through the Stay Slim Eating method. Another asset is that you can't fool yourself into thinking that you're eating fewer calories than you actually are.

There are so many new, low-calorie packaged foods appearing that you can now find them mixed in with the general foods. It pays to look for them in every department instead of, as formerly, in the 'Diet Foods Section'. It's not unusual, for example, to find low-calorie salad dressings displayed on the same shelves as the regular salad dressings.

There's a considerable difference in the calorie savings in salad dressings alone. A pint of one brand of low-calorie mayonnaise (imitation mayonnaise) contains about 2,300 calories fewer than a pint of regular mayonnaise. In taste tests, most people couldn't tell which brand was the regular mayonnaise and which was the low-calorie type, in spite of the enormous difference in calories. Just as many 'testers' who were sure that they did detect a superiority, actually preferred the low-calorie type as those who selected the regular.

Important: Don't Show the 'Low-Calorie' Package

It's vital to realize that these people 'testing their taste' consider themselves gourmets, not just ordinary eaters. If their highly-developed palates couldn't tell the difference between those low-calorie and regular foods in many categories (true in tests of carbonated beverages, also), it isn't likely that your family can either.

However, when people see the 'low-calorie' label on the food before they taste it, then they're usually positive that the flavour is inferior. This too has been proved time after time. Therefore it's important that you don't show the 'low-calorie' containers. Of course, if you're the cook, you can't fool yourself. However, you are certainly intelligent enough to decide the merit of a food on the basis of its actual taste rather than on prejudice against the flavour of low-calorie foods in general.

Even if the low-calorie food, in your naturally biased opinion, doesn't taste 'quite as good' as the regular, isn't it worthwhile eating something not 'quite as good' to help have a slim and attractive figure? Surely you can enjoy the dietary dressing which has only one to three calories per teaspoonful rather than ten times as many or more. This can make a big difference in preventing ugly pounds from piling on.

Flavours Keep Improving

It pays to keep trying the low-calorie foods even if you've been disappointed before in the past. Food processors are making great strides in improving flavours all the time. It wasn't until the time of this writing that I could find a low-calorie 'mayonnaise' that I felt dieters would really enjoy. Those in the past had tasted flat or otherwise were unsatisfactory to the palate. So don't give up just because you were once disappointed. Keep on searching and tasting. It's worth it in order to help stay slim.

You can always get your money back from a reputable manufacturer by returning the product if you don't find it satisfactory. Don't hesitate to do this. For one thing, it's the quickest way for the maker to learn that he'd better improve his flavours or lose money in a hurry. By speaking up you help the food processor, other calorie-conscious people and yourself. What's more, you won't be wasting any money in case you can't use up the product.

Not so long ago the prices of low-calorie foods were much higher than the regular. Today the prices for both types are much more in line. The low-calorie 'mayonnaise' mentioned was only a few cents more than the regular and can be expected to be about the same soon. As the demand for the new type increases, savings can be effected by producing in larger quantities. Under our competitive system such savings will eventually be passed on to you.

Keep checking costs of regular versus low-calorie foods at food markets. More and more you can find comparable good values in canned fruits, vegetables, jellies and many other popular foods. In some cases, you'll learn that you can even save money. Non-fat dry milk costs much less than regular whole milk, for example.

Some of the Best Calorie-Saving Foods

Non-fat instant dry milk has proved to be a great aid to those who watch their weight. You get just as much nutrition as with whole milk since only the butter fat has been removed. But you save half the calories. The flavour of non-fat dry milk has improved considerably in recent years. It will probably become even better in the future.

With this food, as with so many others (as covered elsewhere), many people find that they prefer the flavour of the low-calorie product after a while. I've had many patients tell me that when they tried whole milk later it made them feel a little ill as they found the flavour and aftertaste 'heavy and fatty'.

The same was true with many individuals when they switched from coffee and cream to black coffee with or without artificial sweetening, when they eliminated use of butter or margarine as a spread or melted on vegetables. (I have nothing against those fine foods except for the excess of calories for weight-watchers.)

Most low-calorie beverages, as I've suggested often, are just as tasty as those containing sugar but with practically no calories at all. This development has been one of the greatest helps in reducing. A good, low-calorie carbonated drink is refreshing and satisfying. I often urge dieters to drink as many bottles a day as they wish. Such beverages satisfy the 'sweet tooth', give a sense of filling the stomach temporarily, and are excellent 'instead-of's' when you're about to reach for sweets, cake, biscuit, or sugar-loaded, calorie-heavy beverage.

Always keep a plentiful supply of artificially sweetened minerals around in a variety of flavours, preferably in the refrigerator so you don't even have to go to the trouble of adding ice. You can drink a dozen different flavours in a day without adding a meaningful amount of calories to your day's consumption.

This applies not only to your stay slim eating habits, but also with any quick weight loss diet you use. These drinks have been lifesavers for many of my patients in helping to keep them from breaking a diet for want of a sweet drink, a cake, a biscuit, or another calorie-rich snack.

Hot and iced coffee (except where one or the other is forbidden on a few special situation diets) may be enjoyed as much and as often as you wish. Artificial sweeteners have made these drinks palatable for many who require the sweet taste in a drink. A refreshing glass of iced tea with a squeeze of fresh lemon and with artificial sweetening to taste, is every bit as delicious as 'the real thing', without adding any calories which mount up to pounds of overweight.

Many fruit juices are now, or soon will be, available in lower-calorie, sugar-free form. Keep in mind, however, that you are adding calories when you imbibe such fruit drinks, even though reduced considerably by the absence of added sugar. A typical 16-ounce can of artificially sweetened apple-grape juice is 80 calories total, 20 calories per 4-ounce serving. A 16-ounce bottle or can of artificially sweetened grape soda adds *no* calories to your intake.

Remarkable progress has been made in the artificial sweeteners themselves. You can get them and enjoy them in any form now – tablets, liquid, or crystals which are used by the spoonful like sugar. The old prejudice against sweetening with tablets or liquid instead of with crystals like sugar must necessarily be put aside as an alibi.

As for their safety as a food, thorough investigations to date haven't disclosed any findings that artificial sweeteners are unsafe for young or old. Like my Quick Weight Loss Diet, this advance makes it easier than ever before possible for you to slim down to your ideal weight and stay that way.

Low-calorie 'full course' meals are now on the market, with total calories marked.

Packaged 'low-calorie' types of popular foods include such favourites as:

Bread and crackers in low-calorie versions.

Sugar-free, high-protein cereals.

Artificially sweetened jams, preserves, and other fruit spreads.

Artificially sweetened or no-sugar-added fruits and fruit cocktails. They're delicious and save many calories over fruits packed in the usual heavy, sugary syrups.

Salad dressings now come in a variety of low-calorie types. You can also make your own low-calorie salad dressings using fresh lemon juice, vinegar, dry seasonings, tomato juice, and so on.

Desserts are available in many low-calorie forms. Artificially sweetened jelly comes in many flavours, indistinguishable from the 'real thing'. Low-calorie whips from a package or pressurized can look like and substitute well for calorie-heavy whipped cream. And, of course, what is more refreshing for dessert than naturally low-calorie melon, with or without a touch of sherbet?

'From low-calorie soup to cheese' has now become a reality. Fat-free consomme and bouillon contain no calories to speak of. They may be enjoyed at meals and between meals instead of a snack loaded with calories. The hot or cold consomme will usually give you greater nutritional benefit too. There are various calorie-reduced cheeses including skim-milk cheeses, cheddar, and cream cheese types, as well as delicious natural creamed cottage cheese.

Artificially sweetened sweets and chewing gum provide something sweet and satisfying when your appetite craves it, with very few calories added. Try the mints, fruit drops, and other types such as gumdrops.

New Food-Processing Methods Preserve Flavour

If you were able to understand and know in detail the food-processing methods by which 'low-calorie' foods are made, you'd realize that the quality can be as high and the products as 'safe' and usually as nutritious as the higher-calorie processed foods. New techniques have been developed to take out much of the fat from meats by hand-trimming them first, then subjecting them to high-heat processing which removes the fat quickly and efficiently with a minimum of flavour loss.

Instead of oil in salad dressings, non-caloric vegetable colloids are used. Sauces are made with the aid of non-fat solids and low-calorie vegetable extenders instead of creams, starches, and other high-calorie ingredients. High-calorie sugars are replaced by non-caloric sweeteners in making many dessert-type foods.

Natural shortcuts are used whenever possible, most of them being just good sense and ingenuity. In making low-calorie chicken dishes, for example, often only the white meat is used instead of dark meat which is higher in calories. The fatty skin is removed first.

New kinds of equipment have been devised. For making low-calorie soups and salad dressings, large, high-speed centrifuges are utilized. They whirl the fat to the surface. There it is quickly and thoroughly removed by special cooling and separation methods.

To add extra taste which may have been lost in the removal of fats, carefully selected extra-tasty vegetables, herbs, and flavourings are used (more than may be common in high-calorie processed foods). Favourite ingredients include onions, garlic, spices, herbs, and Italian tomatoes, all low in calories but high in the flavours that often distinguish gourmet foods. Wines and wine flavourings are used judiciously in order to add fine aroma and taste without many extra calories. All these excellent devices, long used by the 'best cooks in town', help add delicious taste but very few calories.

In dressings, non-nutritive vegetable stabilizers are used frequently in place of heavy-calorie oils. That's why an ounce of low-calorie dressing may be only 15 calories compared with up to 150 calories per ounce of a similar 'regular' dressing. Certainly that's a worthwhile saving when there's no vital loss of flavour.

Non-fat milk solids, a fine nutritive ingredient, are used as tasteful substitutes in many recipes, again bringing down the calorie totals considerably.

In lots of cases, although the price may necessarily be somewhat higher, the low-calorie product can taste better than the 'regular' item. This is true, for example, with some preserves. A top-quality, low-calorie preserve on the market is made with 80 per cent natural fruits, plus non-caloric sweeteners. Many 'regular' preserves average about 45 per cent fruit and 55 per cent high-calorie sugar or other stretchers. Wouldn't you rather enjoy the low-calorie, almost-100 per cent fruit preserves as you get the benefits of fewer calories?

As one case history of how helpful low-calorie foods can be, here are the results of a test conducted at a hospital under careful supervision. Fifty overweight men and women were told to eat the same quantities as they had in the past (not more) but substituting low-calorie foods for the regular foods.

At the end of a month, the average weight loss was 14 pounds. The doctor in charge of the test said that there were very few complaints about the quality and flavour of the low-calorie foods. He noted that the patients were well pleased with their meals, and especially with the drop in weight.

Important Warning on Low-Calorie Foods

While lots of the new foods save you many calories, you still must keep counting calories when eating them or your daily intake can go above the point where you take off weight. For example, a patient told me virtuously that she couldn't understand why she was gaining weight again after taking off many pounds on the Quick Weight Loss Diet. She said that on her Stay Slim Eating plan she was eating a number of 'low-calorie' foods.

Upon questioning, I learned that she was eating loads of low-calorie foods *without counting the total calories*. For example, at lunch she'd have a whole 13¼-oz. can of low-calorie chicken à la king. When she brought in a duplicate can at my request, I showed her that the instructions suggested 'three-ounce servings of 72 calories total.' By eating the whole can at a meal, she was absorbing 324 calories. When she added vegetables and other foods, including too much 'low-calorie pudding' on top of this, she was going way above her Stay Slim Eating allowance.

She learned her lesson, one that you must keep in mind for the rest of your life. She went back on the Quick Weight Loss Diet for a week which brought her down to her ideal weight again. She returned to Stay Slim Eating, using low-calorie foods sensibly. She has had little difficulty in maintaining her attractive slimmed-down figure since then.

Just because a package is labelled 'low-calorie', 'lower in calories', 'sugar-free', 'artificially sweetened', or some other such phrasing, gives you no licence to eat limitless quantities. If you're not always careful you won't stay slim. Check the exact measurement in calories printed on the package.

Don't be fooled by 'calories per three-ounce portion', for example, if such a portion of the food amounts to practically nothing in bulk and is not in fact even a moderate portion. Go by the *total calorie intake* involved. That's the only sure way to measure the number of calories you're adding at each meal.

Count Calories in Preparing Servings

Your own intelligence and creativity can save many calories for yourself and your family in the preparation of foods. For example, you'll find that many recipes which call for whole milk can be made with skim milk and are just as delicious – some people say they're even better. You can mix up as much skim milk as you need at the time, using non-fat instant dry milk. That saves you from keeping a bottle of milk handy and perhaps wasting money when it goes sour. More important, the skim milk cuts out half the calories compared with whole milk in the recipe.

You can keep finding more calorie-saving cooking tips all the time in women's magazines and recipe books. For example, if you like sour cream, which is as high in calories as real cream, here's how you can make a delicious substitute. Whip some creamed cottage cheese in a blender to a smooth, sour-creamy consistency. It's satisfying but with only a fraction of the original calorie count.

To prove that a gourmet meal need not be high in calories, consider this dinner menu:

Consommé with herbs, hot or jellied
Half a spring chicken, roasted in its own juices to a golden
    brown (no butter or margarine needed)
Asparagus with low-calorie 'Hollandaise'
Endive salad with low-calorie Roquefort dressing

Minted melon balls, iced
Espresso coffee with artificial sweetener

For your own pleasant surprise, total up the calories
according to the earlier calorie listings for this fine dinner.
You'll see that this becomes a low-calorie meal for your Stay
Slim Eating. Here are just a few recipes, chosen from many
by food experts, to prove to you that low-calorie servings can
be delicious too:

Grilled half-grapefruit with mint garnish – 75 calories

Melon-balls-and-strawberries, soaked in their natural juices
and chilled thoroughly – half-cup – 55 calories

Chilled melon, moderate-sized wedge, garnished lightly
with artificially-sweetened mint jelly – 80 calories

Grilled banana, first lightly spread with artificially sweetened strawberry jam – medium size – 90 calories

Sherried orange sections, one orange sectioned, lightly soaked
in sweet sherry and iced – 40 calories

Eggs Arabesque, omelet of one egg, half tomato, ⅓ green
pepper, herbs and spices – 110 calories

Chicken livers *en brochette,* 3 livers, 1 slice bacon, 4 mush-
room caps – 165 calories

Majorcan fish casserole, 1 slice fish, moderate portion rice,
tomatoes, green pepper, saffron, in white wine – 220
calories

Chinese Delight omelet, 1 egg, 1 teaspoon each of mushrooms, celery, onions, green peppers, in no-fat frying
pan – 100 calories

Chicken Divan, 3 slices breast of chicken, ½ cup broccoli
seasoned with herbs, 1 tablespoon Parmesan cheese – 230
calories

Stuffed baked potato, small size, stuffed with 3 tablespoons of cottage cheese mixed and garnished with
chives – 90 calories

Ginger-baked carrots, 3 small carrots sprinkled with ginger
and baked until tender – 50 calories

Asparagus salad, 4 spears of asparagus, green pepper ring, low-calorie French or Italian dressing – 85 calories

Orange-ambrosia Cup, half fresh orange sectioned, soaked in light brandy, chilled on ice – 75 calories

Orange-grapefruit Royale, to fill medium-sized dessert glass, mix orange and grapefruit sections with small cubes of artificially sweetened jelly, garnished with small orange slice – 45 calories

Springtime Fruit Cup, fill medium-sized dessert glass with sliced fresh strawberries, diced fresh pineapple, fresh grapefruit sections, soaked in artificially sweetened lemon juice, garnished with mint leaf – 45 calories

Zero-Calorie Servings

Here are a few varied servings which show that some delicious foods can be eaten without consuming enough calories to even count them, such as:

Russian Borscht, dissolve 4 bouillon cubes in 4 cups of boiling water, cook with half-cup grated raw beetroot in covered saucepan 15 minutes, cool and chill, add 2 tbsps lemon juice just before serving, salt and pepper to taste, garnish with cucumber slice

Tomato-celery soup, dissolve 2 bouillon cubes in 2 cups of boiling water, in medium-sized saucepan with 1 cup of finely diced celery, ¼ cup finely chopped onion, 1 cup tomato juice, salt and pepper to taste, simmer until vegetables are tender, serve hot

Mint sauce, combine 2 tbsps water, ½ cup vinegar, ½ tsp liquid no-calorie sweetener, leaves from 1 dozen sprigs of fresh mint finely chopped, let stand in warm place for about an hour

These recipes are just a few of hundreds you can make with or without processed low-calorie foods to serve tasteful and satisfying low-calorie meals. Just a little extra care, and an attitude that Stay Slim Eating can also be enjoyable eating, can help keep health high and figures trim for you and your entire family.

If you become careless and the pounds start creeping up on you once more, be sure to go back on the Quick Weight Loss Diet or your choice of the other quick reducing diets.

15. Vital Reducing Advice

Here is additional important and helpful information on some vital subjects only touched on previously.

Dieting for Teenagers and Children

This book is written specifically to help overweight adults. The recommendations are therefore not necessarily for teenagers and younger children, though I have helped hundreds of youngsters over a wide age range to reduce. I consider overweight at any age either immediately or potentially dangerous.

Unfortunately overweight parents are usually responsible for their children being fat. Studies show that when one parent is overweight, there is up to a 50 per cent chance that the child will be too – and up to an 80 per cent chance if both parents are heavies. When a mother and father are grossly overweight and try to get the child to reduce, it's little wonder that chances for success are small.

Children are primarily realists who believe, 'Don't just tell me, show me.' If you don't lose weight, there's far less chance that your child will, although some youngsters, especially when in their teens, exhibit more willpower than their parents.

Being overweight should not be considered a minor problem for a child. Surveys have shown that overweight children are generally more dependent to an undesirable degree. They have more school and sex problems, and are more often emotionally disturbed than youngsters of normal weight.

Despite old-fashioned ideas to the contrary, the fat child is not usually a healthier one. A leading pediatrician stated, 'We don't want babies to be fat any more.' If your child is overweight I urge you to discuss it thoroughly with your physician.

Don't adopt the too-common attitude, 'He (or she) will grow out of it.'

Your doctor will determine whether or not your child is one of the tiny minority (usually noted as less than 5 per cent or one out of twenty) whose overweight is caused by some metabolic, glandular, or other disturbance. Too often that fat child becomes a fat adult with all the great dangers involved.

Keep in mind that serious disorders adversely affected by overweight – including high blood pressure, diabetes, and hardening of the arteries – may be started in the teens and perhaps even younger. There's no doubt that the spurt of growth during adolescence increases the body's nutritional needs, but when excess fat piles up that's a danger signal not to be ignored or misinterpreted as 'nutritional need'.

In dieting, children and teenagers should be watched more closely for vitamin/mineral deficiencies although these are not likely to occur. The youngsters who are growing need more protein, calcium, and iron than most adults, but properly planned food restrictions and increased exercise are often essential. Overweights, young or older, are usually less active, more sedentary and lethargic.

In a conference at the New York Academy of Medicine, it was stated that 15 per cent of high school students are overweight (that's based on 10 per cent or more over average weight). In cases where both parents were heavies, so were 73 per cent of their offspring in this report. If you are a similar 'case', I urge you strongly to do something at once to bring down your child's weight and your own. When both diet together, there's a better chance for both to succeed.

Dieting for Elderly Persons

Younger people can generally look to the elderly as good examples of losing weight rapidly when necessary, proving that it can be done. In my experience, when elderly persons are told to reduce because of coronary heart disease, peripheral vascular disease, generalized hardening of the arteries, or for other reasons, they have little trouble staying on strict reducing diets. (If they had reduced earlier in life, most of them would not be afflicted so seriously now.)

These people know that this is their last chance to save themselves. The desire to live is strong in them and provides powerful motivation. They don't have to think about what may happen to them in twenty or thirty years, the problems and challenge are here and now. They will usually follow almost any necessary regimen. They rarely break training when placed on a diet.

It has been recommended by the American Heart Association's Nutrition Committee that overweight can be prevented by cutting food consumption 1 per cent every year after the age of twenty-five. Instead of growing heavier as one grows older, as is common, the opposite should be true for better health and longer life. The elderly are generally less active and therefore not only should, but must, eat less if they wish to survive the menace to health brought on by forcing the heart to push around excess poundage.

Most individuals should remain at the same weight that they were in their upper teens (unless, of course, they were overweight then). Physiologically after those years the normal metabolism of energy usage by the body remains stationary. The amount of calories or energy necessary to sustain life and health decreases. But as individuals grow older and more prosperous they tend to eat more and get about less. Between the food intake increasing and the energy expenditure decreasing, an imbalance usually develops which piles up the calories on the plus side. The difference per day or week or month may be slight but as time goes on, the pound or two or five add up and total twenty to fifty or more extra pounds deposited over the years. This insidious slow deposit is the cause of most overweight.

Unfortunately these same persons cannot take off the weight gradually. They usually fail if they try. The elderly man or woman in imminent trouble because of excess fat cannot afford to wait for slow, gradual reducing even if it could work for them. Stringent dieting techniques, preferably under the care of a physician, will reduce the elderly patient as well as others swiftly, as required.

Advice on Eating when Travelling

The advice here primarily concerns eating while travelling abroad. Many patients who visit foreign countries return home either heavier, often to a considerable extent, or sick. It's vital in trying to avoid sickness of the stomach to understand the usual reasons why this occurs.

The diet of many parts of the world is given to an abundance of rich gravies, sauces, and concoctions which mask the taste, and perhaps the toughness in the case of meats, of the original foods. Too often the fats or butter used in these sauces have been exposed for hours in a warmth where bacteria multiply very rapidly.

Such bacteria are often accidentally introduced into foods by careless cooks and handlers. It is a common practice in many places for the cook to taste the food frequently with a spoon or other utensil. This is reinserted without washing into the same mixture, perhaps time after time, communicating harmful bacteria if existent in that person.

Meanwhile, with the same general careless attitude, bacteria from a cold or sore throat shower the food via coughing and sneezing. This is less likely to happen in the 'better' restaurants and hotels, but it is the desire and privilege of the traveller to seek out smaller and perhaps less careful eating places.

Good refrigeration which helps protect food from contamination is less likely to be found outside of the 'best' hotels and restaurants. When foods cooked in heavy fats are eaten they are apt to upset the stomach, as would happen to the traveller who would eat too much fat at one time at home. Fats may often be overheated; this produces products of decomposition and often the result is indigestion, sometimes very severe.

Furthermore, travellers going abroad must be cautioned not to indulge heavily in foods to which they are unaccustomed.

Spices, herbs, and condiments are apt to be used more exten-sively. These tantalizing cooking tricks are employed to whet the appetite. People on holiday, tasting the enticing foods, usually eat far more than the body requires or can handle comfortably.

The result very frequently is painful and uncomfortable gastritis. If a few alcoholic drinks are added – cocktails plus wines or beer – there is an even greater chance of further inflaming the irritated, oppressed stomach.

It must be taken into account when travelling that rich, creamy foods such as custards and pastry fillings, especially when not properly refrigerated, are real culture media for staphylococcus germs. Food of this nature, particularly if exposed unrefrigerated for a few hours, not only encourages the germs to multiply rapidly but may produce enterotoxin which is very irritating for the gastric mucosa.

All travellers, especially those who are overweight, are cautioned to observe the following basic rules as much as poss-ible to avoid stomach upsets which may take severe and even serious forms:

1. Try very hard not to overstuff at any time. This gorging tends to aggravate the slightest undesirable condition that may exist. No matter how delicious a serving may be, it can be savoured in a small or reasonable quantity instead of a huge overloading.

2. Drink only bottled water in most places.

3. Try to omit all rich gravies and sauces, heavy dressings and mayonnaise.

4. In most places it is much safer to bypass custards and rich, creamy pastries.

5. Though it may be difficult, try to be sure that the water for tea and coffee has really been boiled. This is particularly essential when ordering iced tea or coffee.

6. Avoid cold soups in many places, particularly cold creamed soups.

7. Be wary of any soft, cold concoctions.

8. Rare and medium rare foods in many places are dangerous, more so than well done foods.

9. Eat only when hungry – don't stuff a meal into yourself just because it's 'mealtime'. Such overstuffing can trigger upsets which might not have occurred otherwise.

10. In eating fruits and raw vegetables make sure, if possible, that they are thoroughly washed. It is better to avoid them.

11. In choosing from a menu it is generally safer to seek out those foods which are well boiled, baked, or grilled.

12. If you feel ill, it can be very helpful to be carrying packets of instant porridge, bouillon cubes, whatever non-fatty, bland foods you can make just by adding *boiling* water.

It is difficult and perhaps at times impossible to diet, or to want to, when travelling – especially on vacation. The following menu is given as a relatively 'safe' day's eating, particularly if you are starting to feel at all queasy or uncertain or if your stomach has given any symptoms of upset:

Breakfast:
    grapefruit or melon
    1 hard-boiled egg
    1 slice of toast
    hot coffee or tea (was the water boiled?)

Lunch:
    clean, well-washed, tossed salad with vinegar and a little
        oil only as dressing, or squeezed lemon
    1 slice of toast
    1 portion of cheese
    black coffee or tea or bottled water

Dinner:
    grilled steak, well done
    well-washed sliced tomato
    1 piece of well-washed fruit
    black coffee or tea

With this type of eating you may not enjoy yourself as much as some of the lucky ones who are not badly affected. You will come back without dysentery, typhoid brucellosis, amoebic dysentery, salmonellosis, or attacks of acute indigestion. You will also return without having gained any weight or perhaps you will even have lost a few pounds. At the very least, *keep all the precautions in mind.* Bon voyage!

Smoking and Overweight

Doctors all over the United States hear statements like this from patients every day:

'Okay, Doctor, I'm finally scared enough of getting lung cancer from smoking cigarettes that I want to quit. But last time I tried, I gained 15 pounds in no time even though I was already overweight. And I'm just as afraid of killing myself from being too fat. How can I stop smoking without putting on weight?'

When cigarette smoking is stopped, weight usually goes up. Yet it's possible to keep your weight down even when you stop smoking.

One man told me, 'My doctor says he'll give me another month to take off the 18 pounds I put on since I stopped smoking, or else he'll put me back on cigarettes. He says that if it has to be one or the other with me, then overweight is more dangerous to my health than smoking.'

That's true, as proved by facts and statistics. As you probably know, compared with non-smokers, average men smokers of cigarettes have about ten times more chance of developing lung cancer. Heavy smokers are twenty times more likely to be stricken by lung cancer, along with other dangerous disorders. Smokers on an average die younger than non-smokers from many ills.

Overweight contributes to many disorders that are up to fifteen times more serious than the possibility of getting lung cancer from smoking cigarettes. There's no question that you have a far better chance to live a longer, healthier, and more vigorous lifetime if you slim down and don't smoke.

Here's why the smoker, man or woman, tends to put on weight when he stops smoking. He is usually a person who either by temperament or because he has so conditioned himself, seeks to satisfy every oral desire. As a smoker, he contents himself with cigarettes during his waking hours.

If he no longer has cigarettes, he usually seeks a substitute. He may take to eating sweets, chewing gum, nibbling on fat-producing and rich foods all day and evening. The pounds pile on and weight goes up shockingly.

A cigarette can be an appetite-curber in many ways. Some smokers find contentment in a cup of coffee and a cigarette. At coffee-breaks, the cigarette substitutes for a doughnut or pastry. At meals, many smokers take a cigarette instead of dessert. Some are happy with only a cigarette and coffee for breakfast. Thus a cigarette replaces some food hunger for them. But take away the cigarette and they tend to compensate for the loss of smoking habit and enjoyment by stuffing themselves with foods loaded with calories.

From a physiological or medical viewpoint, a cigarette may act as a weight-reducer in its effect on the system. This is true between meals but not with a cigarette smoked right after a meal, as explained later. Experiments have shown that nicotine works internally to reduce the extent that foods add to fat. When the smoker stops, the fat-inhibiting action halts too, usually accelerating weight gain.

Heavy smoking dulls, almost anaesthetizes, the sense of taste. When smoking is stopped, foods taste better to the individual so that he or she is likely to eat far more and bigger portions, boosting the daily intake of calories considerably.

Here are specific steps to help keep you from gaining weight when you cut down on smoking or give it up completely:

1. Unless you're a supremely strong individual, it's practically impossible to go on a diet at the same time as you stop smoking cigarettes, and stick to either or both. In my practice the worst failures have been those patients who insisted on

denying themselves both the foods and cigarettes they enjoyed so much, all at one time.

If you're overweight to start, go on the Quick Loss Diet at once to get down to your desired weight in a hurry. Meanwhile try to cut down on your smoking gradually over a period of weeks while you return to ideal weight. At that point you'll be better able to reduce smoking drastically or stop completely.

I've let my patients who are reducing smoke up to ten cigarettes a day, except those who must stop all smoking immediately because of specific physical impairment. You can have a cigarette before meals as this tends to curb the appetite, and as you wish between meals, trying to space out your daily limit to ten or preferably six.

Don't inhale and don't take more than ten easy puffs per cigarette. Then put out the cigarette and discard it or even keep it in your mouth unlit if you find that helpful. Filter cigarettes are preferable to all-tobacco, they usually contain less nicotine and other substances referred to as 'tars'. Choose filter cigarettes lowest in nicotine and tars. I consider them less harmful even though this may not have been clearly established statistically.

2. *Very important: Don't smoke a cigarette directly after a meal.* It's a curious fact that while a cigarette before a meal or between meals acts to burn up more fats and stimulates the metabolism, something quite the opposite happens directly after a meal. A cigarette in the middle or at the close of a meal induces a greater absorption of fat from the intestinal tract, tending to add more weight to your body.

If you 'must' smoke, wait an hour or two after a meal before lighting a cigarette: if you can't do without a cigarette after the meal, take three or four slow, easy puffs, not inhaling, then put out the cigarette, holding it in your hand or mouth unlit if that's helpful.

3. If, like a small percentage of smokers and eaters who cut down, you experience 'withdrawal symptoms', don't be frightened by them as possible deep-rooted physical ills. Such persons may develop some giddiness, exhaustion, headaches,

drowsiness, restlessness, insomnia, and a general feeling of uneasiness. An examination by your doctor is the best way to assure yourself that these are expressions of emotional conflict and not physical ills. He may prescribe tranquillizers or other measures.

Until you see your doctor, you may take available anti-histaminics, 'sleeping pills' which have been passed for non-prescription sale, two buffered aspirin tablets four times daily, common fizzy alkalizers, or just warm skim milk to help calm you.

4. After your weight is normal or near-normal, act to cut out smoking completely. It's not easy for most, so try every possible aid like the following to find what will work for you, trying one after the other or combinations if necessary.

Your doctor can help you with means such as injections of nicotine or nicotine-like substances to decrease or modify the craving for tobacco. He can prescribe tranquillizers and appetite suppressing drugs or other medications such as astringent mouthwashes. In difficult cases the doctor should be seen at least weekly for his moral as well as medical support.

If you find it impossible to stop smoking completely, cut drastically to a 'maintenance cigarette diet' of six cigarettes a day as there is little likelihood that this will be harmful. You might try switching to little cigars (as even some women are doing), regular cigars, or a pipe. You're less likely to inhale since the stronger smoke is more alkaline and more irritating to the throat. Most cigar and pipe smokers tend to avoid the harsh irritation caused quickly by inhaling. (Inhaling cigar or pipe tobacco smoke is at least as bad for you as cigarettes.)

If you feel depressed as you quit smoking, drink strong coffee, tea, or non-caloric carbonated beverages more often than usual, for a lift that has proved helpful for some people, without adding calories. You may be aided considerably by use of readily available anti-smoking tablets, lozenges, medicated chewing gum, or other products, some of which contain a nicotine-like substance. Some people are helped by drawing on smokeless cigarettes made of plastic or other materials.

Try one after another of these suggestions until you find the one or combination that will help you stop smoking. It may be a tough fight but if you can cut out smoking cigarettes for two full days, usually the desire for another cigarette starts to diminish and then continues to lessen each day you abstain. The bright day finally comes when you can recall smugly, 'Once I was a heavy smoker . . .'

After you take off the undesirable extra poundage quickly by the Quick Loss method, maintain your ideal weight as explained previously. You'll be much healthier for having overcome the dual dangerous problems of smoking and over-weight.

Cholesterol and Polyunsaturates in Reducing Diets

As knowledge has increased in the study of fats, involving the concentration of public attention on the potent words 'cholesterol' and 'polyunsaturates', the problem has become more rather than less complex for the experts, let alone for you.

Cholesterol is an essential fatty material which exists to some extent in the blood. It is not a fat in itself but is found associated with true fats in the animal body. The relationship between the availability of cholesterol and the mobilization of fats in body depots is an important one.

Too much cholesterol is bad for you; so is too little. If your doctor finds that you have a high blood cholesterol, he will restrict certain foods and recommend others for your case.

Since there is so much not determined definitely on the subject of cholesterol and polyunsaturates at this writing, certainly you cannot know how much each food affects you according to its cholesterol content. It is not even clear scientifically how much of the cholesterol in a food is absorbed by the body.

Your prime concern should not be the cholesterol content of foods or the lack of abundance of polyunsaturates, but rather following effective dieting procedures as recommended here.

You'll be healthier when you're rid of excess fat, whatever causes it.

To clarify some terms: saturated fats come from animal products chiefly (butter, lard, suet, certain oils), usually solidify at room temperature or lower, and tend to raise cholesterol levels. Unsaturated fats come principally from vegetable fats and are usually liquid at room temperature. Polyunsaturated fats have more hydrogens missing than usual unsaturated fats and they are more active in reducing blood cholesterol (many vegetable oils are high in polyunsaturates, but this is not true of coconut oil, olive oil, or cocoa fat).

There seems little question that an elevated blood cholesterol level indicates that a person is more likely to have a heart attack than if he had a lower cholesterol level. It has not been proved conclusively that lowering the higher cholesterol level will specifically prevent heart disease, although indications point that way. Meanwhile I advise overweights as a whole to abstain from or reduce their intake of animal fats and high-fat dairy products in general.

In essence, I urge you to take steps now to get rid of your excess weight, while you let the scientists work out the facts in their area. The one point all agree on is that *overweights should reduce*.

Should Drugs be used in Dieting?

Used correctly under the supervision of a physician only, drugs can be an excellent aid in reducing. So-called 'reducing drugs' (often just vitamins) sold over the counter without a doctor's prescription, or not specifically designated for you by a physician, should never be taken for dieting. Drugs are seldom required for the person who has a compelling and driving desire to reduce. Some individuals need every means available to assist them in their reducing. For them the physician has many good drugs to choose from for various purposes. He alone knows when to give drugs, when to increase or diminish dosage, when to stop their use. The giving of drugs is a very intimate and skilful art. It cannot be delegated to the lay person, nurse, dietician, or druggist.

I have prescribed drugs for patients, including some of the cases cited in this book, with fine effect. I've always checked usage closely and have had only benefits, no harmful results. I deplore the attitude of too many professionals and others who say flatly, 'The use of drugs in reducing is bad' – usually because of cases of improper usage. That's like saying, 'The use of autos is bad and should be abolished,' because some people drive improperly and accidents occur.

Of the many drugs helpful in reducing I've found the appetite depressants and suppressants most valuable. (Technically, for the curious, most reducing drugs may be classified under anorectics, belladonna preparations, diuretics, hormones, intestinal bulk producers, sedatives and tranquillizers, and various combinations.) In my experience, drugs I've prescribed as reducing aids are not habit-forming for 90 per cent of the patients using them. Many who start with them drop their use as soon as they become accustomed to the dieting regulations.

Drugs are useful in reducing the appetite, counteracting mental depression and physical lethargy, increasing mental interest and physical activity, ridding the body of excess salts and liquid and many other functions.

Leave the knowledge and responsibility of drug usage to your physician. Don't worry about relationship to strange terms you may have come across in respect to reducing, such as 'appestat ... thermostat ... automatic appetite control centres' in the brain or elsewhere in the body. Such words usually serve only to confuse.

Your job is to follow all dieting instructions explicitly, especially when taking drugs. Don't juggle your medicines as suits your fancy. Don't increase or decrease dosage on your own. If you experience worrisome symptoms, inform your physician at once.

Concentrate on your dieting and on continuing to lose dangerous excess weight. Don't become upset by your own fears or any admonitions from others if your physician prescribes drugs as reducing aids. As stated in the Journal of the American Medical Association, 'Drugs have a decided role in the treatment of obesity, especially in the first four to eight weeks.'

'See Your Doctor?'

Of course, as advised on other pages, it's best to see your doctor before and during dieting. He will check your physical condition thoroughly and inform you accordingly, including use of drugs or not. My recommendations here are for people without serious disorders such as diabetes, hyperthyroid, Addison's disease, gastric or intestinal ulcers, cirrhosis of the liver, malignancy, tuberculosis, ulcerative colitis, or neurotics for whom dieting creates added problems. Persons with any such disorders should go on a reducing diet only under the close and constant supervision of a qualified physician. And whether on my diets or any others, if you feel any severe, unusual symptoms at the start or at any time, stop the dieting procedure and see your doctor.

However I must emphasize this: it is more important for most of those with diabetes and other serious disorders to get down to the ideal weight and to stay slim than for others. Your own doctor will confirm this.

In reality, few people with the serious medical disorders mentioned are fat. Before diabetics reach a serious stage, they may be heavy, then they are forced to reduce. The great majority are rarely overweight because the severe illnesses have already caused them to reduce either because of lack of appetite or on a doctor's deadly serious orders. It's significant too that overweights with such disorders have usually reduced on what are really quick action diets like mine, of the doctor's choosing.

If you're suffering from diabetes or another serious medical disorder and are overweight, you may be able to use diets in this book under your doctor's supervision, as has been done with many of my patients. I've had excellent results with diabetics on the Quick Weight Loss Diet, with gall-bladder

patients on a fat-free diet, with hypertensives on many disciplined diets which help reduce the blood pressure effectively.

We must be sensible about the advice to 'see a doctor'. Most
overweights won't go to the doctor about reducing. They don't
consider obesity the deadly sickness that it is. Actually if
23,000,000 or more overweights swarmed into doctor's offices,
the physicians wouldn't have time to take care of other ills.

Medical journals have noted that if the usual advice were
observed such as, 'If you have a heavy cold, see a doctor' –
then physicians could take care of little else. Americans average four to five colds a year. If only one a year was a heavy
cold, this would mean some 185,000,000 doctor visits for colds
alone.

Preferably with your doctor's checkup, the prime objective
for people without serious ailments must be to get those excess
pounds off, not just for improved appearance but to prevent
tragic afflictions due to overweight.

No Miracles in Health Foods and Fads

Much dieting advice is doled out by the 'nature boys', the
'naturopathic advisers' and other such self-styled nutritionists,
lecturers, and 'natural food' advisers. Most of these self-
appointed experts insist that the inherent qualities of life,
health, and beauty can be found only in a return to the primitive
in food, clothing, love, or variations on that theme. Only
the 'natural foods' they recommend are supposed to be good
for you, all deviations are bad.

If these people wish to delude themselves that wheat germ,
blackstrap molasses, brewer's yeast, seaweed, liquorice, yogurt,
grapes, and other hallowed items have special health-giving
and beautifying qualities, that is their privilege. *You must
realize, beyond question, that such foods have no greater quality
for sustaining life than other natural, processed, or concentrated
foods have which are readily available and usually far less
costly.*

Some of these advisers back the eating of only vegetables,

or only fruits, or meats and salads and no other foods, or other systems. Many of the recommendations are excellent ways of reducing. They usually eliminate the rich calories in pies, pastry, ice-cream and heavy dressings. One of the leaders who speaks and writes of 'wonder foods and diets' fundamentally specifies a meal consisting of a salad, meat, some fruit as dessert, and black coffee. This comprises a fine reducing diet, no matter who advises it. There is no danger in this diet, as there is little likelihood of damage from the most lopsided diets if they are not extended over a long period of time.

The real danger is not so much that these 'health foods' usually cost much more and thus in effect cheat you but that they undermine confidence in excellent staple foods. Far worse, the faulty preachings may keep people from seeing a physician for proper treatment of arthritis, cancer, and other serious ills. The health foods are regarded as cure-alls by too many gullible people.

The statistics again are the best proof that a return to the 'old-fashioned ways of eating and living' would be damaging and dangerous rather than helpful. In 1900 the average life expectancy was under fifty years of age; today it is seventy or over. The American Medical Association's Council on Food and Nutrition states, 'There are no *health* foods.' The usual foods available in most food stores are healthful for the vast majority of persons when not overeaten to the point of piling up deposits of excess fat.

Sexual Potency and Overweight

'Does sexual potency increase as overweight diminishes?' I've been asked that question hundreds of times. The answer is necessarily unclear because so much of the relationship between sex and overweight depends on individual temperament and psychological factors.

I've had some male patients tell me that while they were reducing and after they lost a good deal of weight, their sex desires and physical abilities had increased greatly. Women have told me the same about their husband's actions when slimmed down, along with their personal increased sexual desires. However these reports are not sufficient to provide an average or covering answer.

The sex act itself is not hampered by overweight unless layers of fat inhibit the organs and cause difficulties. The libido and passions are not necessarily disturbed by excess weight, although a general lethargy and lessening activity produced by overweight may affect sexual desires and capacities adversely.

One cannot overlook individual attitudes and cultural mores. Only forty to eighty years ago, most of the beautiful women in America were twenty or more pounds overweight by today's standards. Currently the slim figure is commonly regarded as most attractive and appealing whereas the fat man or woman usually draws a sharply negative 'Ugh!' from the opposite sex. Of course some cultures still glamorize the meaty male as the most desirable sex symbol.

It is my belief that a relatively small increase in weight does not diminish the sexual drive on the average. But as overweight increases, greater fatigue is caused by the sexual act as by general exercises and activities which require physical exertion. Conversely, as the overweight takes off debilitating poundage, his general sense of well-being, vigour, and energy usually increases.

It is a known fact among gynaecologists that a reduction of weight is usually accompanied by increased fertility, but there is certainly no guarantee that this will be so in each case.

You may have read here and there that some overweight men and women are glad and even eager to be fat in order to escape the attentions of the opposite sex, marriage proposals, and involvements. In my practice exactly the opposite has been true. One of the greatest motivations for reducing among my patients has been the desire by women to slim down in order to attract a man for marriage or to re-attract a straying husband – or by men to be more appealing to women or to win back an increasingly indifferent wife.

Among the gratifying comments I've heard repeated many times by patients is a man's, 'I feel like a new man!' and a woman's, 'I feel like a bride again!' and, 'They tell me I look years younger!' If such a result holds high promise for you and decisive motivation for reducing, make the most of it. I cannot promise the dieter greater sexual fulfilment but I can emphasize the incontrovertible statistics that point to more youthful appearance, better health, greater vigour, and longer life when not fat.

Exercise and Weight Loss

Proper exercise is a valuable asset to a reducing programme. A special report by a committee of the American Medical Association stated that there is some evidence to suggest that exercise has a beneficial effect on metabolic functions that combat obesity, in addition to burning calories.

It's desirable to exercise all your life, modifying and slowing down the competitive forms as age progresses. Many exercises are fine at any age: walking, bicycling, gardening, dancing. Choose the form of exercising and sport that appeals most to you and go at it daily for long enough to do you good, but stop at the slightest hint of exhaustion.

When you walk, move briskly, purposefully. If you saunter lazily you won't be getting much exercise. Lift your chest, pull in your abdomen (tell yourself as you walk, 'AbdomIN!') and stride alertly, conscious of the movements of your legs and body. Your muscle tone and posture will benefit.

A very simple exercise to help you flatten your stomach bulge is this. Pull in your abdomen as though you're trying to make it touch your spine, while you breathe naturally and count to yourself slowly to ten. Do this while standing and walking, at least a half dozen times a day. In a month or two your abdomen will be flatter and firmer. Also your posture will be improved.

Exercise is a highly recommended aid, but keep this in mind always: *no amount of exercising will reduce you if you overeat.*

To get down to ideal weight, combine proper exercising with swift reducing on the Quick Weight Loss Diet.

Approximate Calories Used Per Hour by Activities

To help you understand how many calories are used up by the body when you're active rather than inactive, note the listing that follows. The figures are necessarily approximated since each individual exercises, works, or walks with differing energy usage. A 120-pound woman uses about 240 calories per hour when doing active housework such as using a vacuum sweeper. She uses up only about 84 calories per hour when sitting and resting, talking, reading, knitting, and so on.

	120 lb. woman	160 lb. man	per lb.
Bicycling, moderate	192	256	1·6
Bicycling, energetic	480	640	4·0
Cooking, active	96	128	0·8
Dancing, moderate	264	352	2·2
Dressing, undressing	96	128	0·8
Driving auto	108	144	0·9
Eating	84	112	0·7
Exercise, moderate	240	320	2·0
Gardening, active	276	368	2·3
Golf	144	192	1·2
Housework, active	240	320	2·0
Ironing	120	160	1·0
Laundry, moderate	120	160	1·0
Lying at rest	60	80	0·5
Office work, active	120	160	1·0
Painting furniture	144	192	1·2
Piano playing	132	176	1·1
Rowing	600	800	5·0
Running	444	592	3·7
Sawing wood	372	496	3·1
Sewing, knitting	84	112	0·7
Sitting at rest	84	112	0·7
Skating	252	336	2·1

	120 lb. woman	160 lb. man	per lb.
Ski-ing	624	832	5·2
Standing	84	112	0·7
Table Tennis	300	400	2·5
Swimming	480	640	4·0
Tennis	336	448	2·8
Typing	108	144	0·9
Walking moderately	168	224	1·4
Writing	84	112	0·7

You'll realize from such figures that even walking at moderate speed for a full hour, while excellent exercise for your body, uses only about 200 calories. So you still must watch your calorie *consumption* carefully. You can't let yourself go and eat a lot at the next meal just because you exercised for an hour or more.

Exercising to Help Your Figure

There are many different exercises which will help keep your figure trim all over and which can help to firm your abdomen and other parts of the body. Here are a few exercises you may wish to try as a start. To get the most good from any exercising you should do it continuously for fifteen minutes or more and day after day. Proper exercising helps to tone up the system and improve the circulation, as well as aiding towards having a trim physique.

A few cautions are essential about exercising. If your legs tend to be thick and heavy, for example, don't expect any 'miracle' transformations. Proper diet and exercise will help your legs to become as slim and trim as possible but cannot change the basic structure. The same applies to other parts of the body.

Never exercise to the point of exhaustion or feeling any pain. Don't over-exert. If it hurts, stop – it's bad for you; continue later or the next day.

Try all or as many of the following exercises as you wish. Start by doing each or several of the exercises just a few times. Then increase the number of times you perform each exercise day after day.

To help trim figure:

1. Stand erect. Holding arms straight in front, clasp your hands. Very slowly bring arms forward and overhead. In that position, roll head back as far as possible without straining, then roll head to the right, then to the left, then to the front. Slowly bring down arms forward, hands still clasped, to original position. Repeat ten times.

2. Stand erect, arms straight forward. Swing right leg slowly up forward, then far back, then bring to rest on floor as you were. Do the same with the left leg. Repeat five times alternately with each leg.

3. Stand erect, arms relaxed at sides. Bring left knee slowly up to your chest, bending upper torso and head forward, clasping leg to chest with hands. Return to original position. Do the same with the right leg. Repeat five times alternately with each leg.

4. Stand erect with hands on shoulders. Bend forward slowly as far as you can comfortably, holding upper torso straight. Then bend slowly from hips to right, then back, then to left, then forward, then back to original erect position. Now repeat the cycle, starting this time by bending forward, then to the left, back, right, forward, up again. Repeat five times each alternating cycle.

To help slim waist and flatten rear, firming legs and thighs:

1. Stand upright with arms held straight up overhead. Bend forward from hips, letting upper body relax, head and arms hanging loose, not trying to touch the floor. Keep relaxed, bend knees slightly, then holding arms forward return to original upright position, arms straight up. Repeat slowly ten times.

2. Sit back in a straight, hard, armless chair, holding chair

seat with both hands. Raise right leg, hold straight forward, move back and forth about eighteen inches, repeat ten times.

Lower right leg, raise left leg and repeat the same exercise ten times with left leg.

Lean back, hold tight to chair seat, lift both legs together straight forward, then holding both legs tight together move them from side to side about eighteen inches, repeat ten times.

With both legs straight forward, move one leg after another up and down like scissors blades, ten times.

In same position, holding legs straight forward, cross one over another alternately, ten times.

3. Take kneeling position, spread legs apart slightly, thrust arms down between legs, and rest hands on the floor. Moving the knees, 'bounce' your rear up and down slowly ten times.

To help slim and firm legs, also abdomen:

1. Lie flat on back on floor, arms straight out at sides at right angles to torso. Bend right knee up. Keeping knee bent, turn leg to left, trying to touch floor with right knee at left side of body without straining. Return slowly to original position. Go through same action with left knee, trying to touch it to floor at right side of body, then bringing it back to original position. Repeat five times alternating with each leg.

2. Lie flat on back on floor, legs straight forward on floor, arms back over head resting on floor. Bend both knees together up close to stomach and clasp rather tightly, but without strain, with both hands. Return arms slowly overhead as you straighten both legs together up towards the ceiling, then lower them together, held straight, to the floor. Repeat slowly entire action ten times.

To help slim upper arms:

1. Stand erect, arms down at sides. Raise arms slowly upwards by bringing straight out at each side until palms are flat against each other overhead. Now face palms outward and keep arms straight, return in sideward motion slowly to straight down. Repeat ten times.

2. Stand erect, arms at sides. Raise arms straight over-head, palms facing each other and about twelve inches apart. Turn palms towards back wall. Make tight fists. Pull arms down slowly, bending elbows, as if pulling weighted ropes down from ceiling, until fists are at shoulder height. With fists still clenched, raise arms slowly as if pushing weight up until arms are straight. Repeat up-down action slowly ten times.

To help tighten under-chin: Sit well back in a hard, straight, armless chair. Let arms hang down loose at sides. Sit straight, shoulders back, but not tense. Move head as far back as you can; in that position open mouth wide, then close slowly. Open and close mouth slowly ten times. Do at least three times daily. This also helps to relax you, as does all good, moderate exercising.

Activity Helps Maintain Health and Slimness

No matter how inactive you may have been in the past, or in recent years, you'll find that you'll enjoy activity, exercise, and sports very much once you make them a natural part of your living pattern. Instead of going everywhere by auto, try to make it a habit to walk to your destination if not too distant. Instead of sauntering along, walk briskly with shoulders back and head high. Take strides that require some energy and use up more calories than in strolling.

Instead of hopping into a car without thinking, to go to the shops or post a letter, walk if it's not more than fifteen minutes or so away. Instead of sitting and chatting with a neighbour or a friend, suggest that both of you talk while walking briskly. Instead of hailing a cab to take you to a restaurant not too far away, enjoy the benefits of walking there and back.

As for sports, you don't have to be an athlete to get lots of good from them. As you slim down from the Quick Weight Loss Diet or other effective diets, you'll find that you'll get a lift mentally as well as further health benefits from bicycling, swimming, walking, and other physical activity.

Philosophically, too, you might consider the advice of Diogenes: 'By constant exercise one develops freedom of movement – for virtuous deeds.'

Your Decision: Eat To Live . . . or . . . Live To Eat

I am most hopeful that this book will lead you to take the necessary and effective steps to reduce – starting right now. I urge you, for your health's sake, to use these recommendations which I've developed during my own lifetime in medicine.

As I have done, along with many thousands of patients, use these suggestions to get down to your ideal weight, and then to maintain that reduced weight. The Quick Weight Loss Diet, Combination Plan with Stay Slim Eating and other procedures and tips which have helped so many thousands to reduce successfully can work for you too.

Many of my greatest satisfactions as a doctor have come from watching the obstinately and persistently overweight in my care – particularly those who had tried and failed for years to reduce – finally succeed through my methods. I wish you the same success and the same gratifying results – a more attractive body, greater freedom of movement, more vigorous and sustaining health, and a better chance for a longer, happier, more rewarding life.

The Complete Carbohydrate Counter 40p
with an introduction by Katie Stewart

A quick and easy checklist of Carbohydrate values . . . Mouthwatering
Low-Carbohydrate recipes . . . invaluable to the slimmer and an
essential checking reference for diabetics.

Dr E. K. Lederman
Good Health through Natural Therapy 75p

Convenience foods, lack of exercise, stress, smoking and drinking are
all to blame for the 'diseases of affluence'. With this book – written by a
qualified Harley Street specialist – you can raise your own health
standards: the right diet, the right exercise and proper relaxation. You
will learn how the natural healing of the homeopath and osteopath can
be of real help to everyone.

Dr A. Ward Gardner and Dr Peter J. Roylance
New Essential First Aid 75p

An easy-to-read, copiously illustrated handbook of first aid.

Completely up to date, and written in the light of the tremendous
advances in surgery, and in resuscitation, the book shows clearly and
simply how to carry out the correct actions in the right order.

'The authors have approached the subject of essential first aid in a new,
interesting and sensible way, placing emphasis throughout on life-saving,
speed and commonsense' *Dr A. Lloyd Potter in his Foreword*.

Oliver Gillie
How to Stop Smoking 70p

A smoker loses five and a half minutes of his or her life for every cigarette he or she smokes – according to the latest reports of The Royal College of Physicians. In this book, Oliver Gillie outlines the latest facts on smoking and explains the effective strategies for conquering the habit.

How to Stop Smoking should help you to give up completely. At the very least, you will smoke a great deal less when you have read this book.

Dr Joan Gomez
How Not To Die Young 70p

Are you living dangerously? This book can save your life.

With the aid of 240 practical, multi-choice questions, Dr Gomez indicates the many different actions and dangerous habits which could do your health irrevocable harm, and shows how unnecessary deaths can be avoided.

'The most important book that can be read today for those who wish to live...' DR BERNARD KNIGHT, WESTERN MAIL

'Gives encouraging case-histories of people who suffered from ulcers, coronaries, cancers, etc., and were successfully treated . . . If we all read and acted on this book, doctors would never need to grouse again' SUNDAY TELEGRAPH

Eileen Fowler
Stay Young For Ever 35p
with a preface by Barbara Cartland

Take the easy way to fitness with Eileen Fowler and stay as young as you want to be. Keep your looks, improve your figure, increase your vitality and enjoy living. The celebrated TV and Radio personality shares the secrets of health and beauty with every woman who wants to stay at the peak of her powers.

Peter Blythe
Stress – The Modern Sickness 60p

Why is stress an ever-increasing problem? How does the mind convert
stress into physical illness? When can stress lead to a broken marriage?
Is being overweight a stress symptom? These and many other vital
questions are discussed by Peter Blythe, a practising psychotherapist
and consultant hypnotist who examines every aspect of normal living
and shows where the build up of anxiety-stress-tension plays a
determining part in a variety of illnesses.

Gordon Bourne FRCS FRCOG
Pregnancy £1.75

Having a child can be one of the most exciting and fulfilling experiences
in a woman's life, provided she has the confidence that comes from
knowing exactly what pregnancy involves.

This comprehensive guide is written by Dr Gordon Bourne, Consultant
Obstetrician and Gynaecologist at one of London's leading teaching
hospitals. It provides full information, guidance and reassurance on all
aspects of pregnancy and childbirth. An indispensable aid to the
expectant mother, it will also be of great interest to her husband and
family.

'Sets out in a clear, factual and reassuring way every possible aspect of
pregnancy . . . I would recommend this book to anyone who can buy or
borrow a copy.' MARRIAGE GUIDANCE

Dr David Delvin for the Back Pain Association
You and Your Back 60p

However young and fit you are, it's more than likely that you will one
day suffer back trouble.

Back pain is unpleasant and a nuisance. It costs the country £300,000,000
a year in medical care, sickness benefit and lost production. The cost in
suffering and disability is incalculable.

Dr Delvin explains how your back should function; what sort of back
trouble you could suffer; how to cope with back trouble and how to
prevent it happening to you.

Derek Bowskill and Anthea Linacre
The 'Male' Menopause

'... if you're forty-some, combing hair carefully over the bald spot, getting a bit trendy on ties and shirts, depressed sometimes when sober and pushy with the girls on two or three pints ... then you too, sir, may be showing menopausal symptoms'
YORKSHIRE EVENING POST

Dr Felix Mann
Acupuncture: Cure of Many Diseases £1.00

Today thousands of European and Russian doctors practise acupuncture in conjunction with Western medicine. In this fascinating and illuminating book Dr Felix Mann dispels the mystique of this ancient science and art, explaining in clear and easy terms its origins, theory and practice.

'Let us hope that Dr Felix Mann will succeed in persuading his colleagues and the public that the method produces good results'
ALDOUS HUXLEY

Doctors Penny and Andrew Stanway
Breast is Best 80p

More and more doctors are emphasizing the important advantages of breast feeding for mothers who are anxious to give their child the best possible start for a healthy life. There's immunity from infection, less chance of obesity and less risk of dental decay ... just some of the advantages of natural feeding. This comprehensive guide explains how breast feeding works, how to prepare for it and look after your own health while coping with the rest of the family.